BUSIN
TECHNOLOGY
DEVELOPMENT
STRATEGY
BUNDLE

ARTIFICIAL INTELLIGENCE BUSINESS APPLICATIONS

Artificial Intelligence Marketing and Sales

Applications

Table of Contents

How AI Has been Used in Business So Far

Customer Service Filter

Answering Basic Questions

A good portion of a customer service representative's time is spent dealing with the most basic questions. The problem with the most basic questions is that they are repetitive and have the same answers.

Companies have invested in artificial intelligence as a way of removing the most basic questions needing attention from human customer service representatives. This allows the customer service representatives to handle more serious cases and respond quickly to customers dealing with more problematic situations.

Correct Redirecting

Including the most basic questions, a common reoccurrence for customer service representatives is simply answering where something can be found on the website. When you have a website as large as, maybe, Amazon then you have a lot of areas where content is buried underneath what is usually used by the common customer.

Artificial intelligence can pick up on keywords in the phrase such as "can you help me find" in order to redirect the user to what they need. Like handling the most basic questions, redirecting customers that are simply lost on the website saves customer service representative time.

Offering Products as Solutions

In addition to this, maybe the customer is looking for something that isn't obvious. Along with redirecting the customer, if a customer is looking for a specific product of yours but can't find it then they will able to redirect them to the appropriate product.

Since products usually complement each other in a company environment, this is often the best time to bring up additional software for sale or other products that complement what they were originally looking for. As this happens in almost immediate timing for the customer, the customer is not ready to defend themselves against a sales pitch and so they will usually check out the additional product.

Immediate Gratification

Due to the fact that artificial intelligence can handle as many people as you want it to, that means that waiting lines are significantly reduced. Instead of that frustration building up as they wait 30 minutes on the phone, the artificial intelligence is able to immediately gratify their needs.

In addition to this, artificial intelligence is less likely to make mistakes. Provided the artificial intelligence is built sufficiently, it will be able to handle the customer's needs without misinterpreting the words of the customer. This is a point of contention amongst customer service representatives that primarily deals with demographic locations and how things are said amongst individuals. The artificial intelligence can recognize the pattern of words and determine the meaning based off of a database rather than personal experience.

More Satisfied Customer

By being able to handle customer service issues in a very quick manner, the customer normally leaves the discussion satisfied with the result. In the past, when customers had to wait to be seen by Representatives even though they may have paid a premium price, the lack of instant gratification often led to resentment with certain customer service agencies. One that comes to mind as a particular example is the Comcast customer service line, which was considered to be one of the worst for nearly half a decade simply because of mistakes and the amount of time that you had to wait.

Customers that are satisfied with the customer service experience are less likely to leave and more likely to stay on your platform, which means they are more likely to purchase items on your platform.

Detecting Fraud

Zip Code Detection

One of the easiest artificial intelligence implementations for security fraud is the detection of the common zip code. When a customer makes a purchase, they usually stay within the immediate area that they are in. ZIP codes that are outside of that area are normally very difficult to detect by human beings.

By having an artificial intelligence constantly looking at the account, it is able to detect when a customer makes a purchase that is outside of their zip code. It is then able to make an assessment as to whether the zip code in which the item was purchased is valid given the user's past travel experience. For instance, a person in Cape Coral is very likely to purchase something in Miami every odd once in a while but that same person in Cape Coral is very unlikely to purchase something in Maryland for 500 times the cost. This brings up the next security fraud detection measure put into place by artificial intelligence.

Out-Of-Habit Detection

When it comes to artificial intelligence, things are grouped together in categories. When security specialists talk about out of habit behavior with financial data, they are usually talking about a three-category system. The first category is conservative, the second category is mixed, and the final category is extravagant or whatever word you want to use to describe blowing all your money. You can detect whether a person is conservative or not by looking at the amount of money they make versus the amount of money they spend as well as whether that money goes to bills or luxuries. By being able to understand the regular spending habits of an individual, a person who is normally conservative is not likely to spend something that is 50 to 100 times the amount they would normally spend. However, if they buy tech related items that are usually expensive, but it happens at a very low frequency then the account would likely consider that a tech purchase that is 50 times the amount that they would normally spend is a valid purchase because of the spending habits of the individual.

Falsifying Credentials

For a very long time, gas stations and security buildings have shared one thing in common. There is always a credential check for whenever you want to do something that requires a special privilege. At a gas station, you are required to provide credentials whenever you purchase things like alcohol and cigarettes. Security buildings, this is more likely to deal with access to certain parts of the building as well as items in that building.

The problem with credential checking is that it was often done by the human eye, which is scientifically proven to be one of the most unreliable devices for scientific measurement. However, with the invention of scanners and artificial intelligence, we no longer need to rely on the human eye to make credential checking more secure. In addition to this, these machines can keep up with current law whereas the security professional in the front is likely to lag behind the law for a few months and, sometimes, never actually update until they're caught in the wrong.

Facial Recognition

In addition to credentials that you might find on a card, some businesses have begun using your face as a credential. It's not very widespread because such a credential can be faked provided you have a highly optimized picture in front of the camera, but the technology is getting much more advanced as time goes on.

For instance, artificial intelligence can already detect movement patterns within the face and take those as additional parameters. By knowing whether the eyes are moving, a common human reflex, the artificial intelligence would be able to determine the difference between a picture and a human. Facial recognition is far harder to fake, should we get to the point where it's as good as artificial credential checking, that it will likely replace credential checking as we know it.

Security Threat Analysis

The last bit is more of an ethical issue right now as many people are trying to fight it as companies are attempting to find ways to determine the behavior of an individual. For instance, a person who is likely going to kill someone in the building is going to be in an off mood compared to their regular attitude at the company.

However, there's an additional side to this and that is the average security audit that is done by the individual. Security audits determine the weak points inside of a building. Artificial intelligence is slowly being able to detect the same security holes in a building that a security audit would normally provide but at a much cheaper cost. This is still relatively new technology and it's not quite as common as the previously mentioned artificial intelligence, but it does exist.

User Data Abstraction

Automate Meetings

In the past, we have usually set up meetings on a daily or weekly plan. In fact, one could say that an hour spent on meetings every day amounts to an entire day wasted at the end of the week. The problem is that meetings happen when nobody really needs to have a meeting.

By being able to understand what is going on in the work environment and on the product, meetings can be assigned based on data of current problems needing to be solved. The only real need for a meeting is to ensure that everyone has a collective understanding, but most of everyone in a company has that understanding as they work throughout the week. It's only when something needs to be changed about a work process or a user problem that everyone goes on to their own page. By using user-generated data, one can assess whether there should be a meeting or that time allotted to the meeting should go to work, which saves time and thus money.

Product Failure Predictions

The most common way to detect whether a product is going to fail or not is the amount of negative feedback that a product gets. It is very easy to determine a good product from a bad product when good products get four to five stars and bad products get zero to one stars.

What artificial intelligence can do is look at the feedback given in the reviews to determine whether there is a product that is going to start failing because of unheard complaints. Instead of looking at just the numbers on the rating sheet, artificial intelligence can collect negative commenting versus positive commenting. As negative commenting is more expressive of an issue and is a real evaluation of the product, artificial intelligence can provide a much more accurate depiction of succeeding and failing products.

Customer Service Refinement

A lot of time is wasted in the customer service industry because common problems keep popping up. Every problem that a customer has can usually be fixed in some way and the more time spent in customer service, the less time the customer is buying your products from you.

By assessing product value and the complaints associated with the products, one can make changes to the product that prevent such complaints. By paying attention to the interactions in customer service, one can generally find where common problems are and remove them from the equation. This allows the company to make more money and to waste less money on customer service representatives for handling basic items.

Website Leads Success

Another area that artificial intelligence succeeds at is determining whether your website pages lead customers to the results that you want. While there are common technologies like heat mapping in the world of front-end development and marketing, artificial intelligence can easily quantify which pages are working and which pages are not based on that information. However, it goes a step further by making predictive guesses on what has worked in the past and what is currently failing so that it can predict what your best design would probably be like and so you don't waste time in revisional steps.

Potential Products

The last part of artificial intelligence that can help a company out in is in products that the company would not have thought of beforehand. Oftentimes, ideas for products come out of nowhere or are based on a logical progression of the evolution of another product.

However, artificial intelligence can pick up on lines like "I think I would like to see this" or "it's too bad that you don't have this". Lines like these are usually suggestions by the customer about products that you could provide and make money off of, but they get lost in the customer representative area because it's not normally the responsibility of customer representatives to report ideas to higher management.

Predicting Area Failures

Machinery Repair Cycles

We all know that machines need to be regulated on a constant basis to make sure that they are working as efficiently as they can possibly be, but the estimates we usually give are kind of standard. For instance, a mechanic will tell you that you need to change the oil in your car in the next 3000 miles or six months. However, this standardization is really just a factor of how often you drive a car. It is not taking in the fact that you might only use the car once every 6 months and so, if that's the case, you might not need to have your oil switched out.

Artificial intelligence can pay attention to every single mechanical repair that is done in the company and then utilize readings to determine repair cycle. For instance, if the company printer is out of ink between 35 and 37 days, the artificial intelligence can determine that you need to order a new ink cartridge on a regular basis based off of that.

It might be easy for a single human to make that schedule, but what if everyone has a printer at their desk? What if you had 50 employees with their own printers, all of which were purchased at different times? This becomes a problem for even a team of people to keep track of if it's not kept track of every time you buy a printer. With artificial intelligence, you simply notify it that there is a new printer in the system it needs to keep track of and then it learns what it takes for that printer to have a problem. This was a really basic example, but this can be done with large technology or industrial machines.

Nonoptimal Production

In addition to this, artificial intelligence can detect when a machine is not moving at the rate that it should within the nanosecond. In a factory, you generally know what's going on by extremely basic sensors, that are already built into the system, that have thresholds.

However, if those same sensors were allowed access to artificial intelligence, the recognition of optimal working speed would be within the nanosecond.

The reason why this is important is that noticing a few nanoseconds of change that continuously change for the worse allows for more of a preventable time window for issues that could essentially shut down the entire factory. As an example, a heat gun might be off by a few nano degrees and it is slowly getting worse. Something like this normally has an internal sensor that can go bad because of overheating, but artificial intelligence would tell you that the sensor was bad and, if you installed a separate measure, it would also tell you that the heating gun was getting too hot or too cold thus creating bad products. Instead of producing bad products for, potentially, thousands of units you instead have the ability to lose maybe a lot of 10 before finding out about this issue and taking care of it.

Customer Patterns

Humans are incredibly predictable about what they will and won't do. This means that since you can predict human behavior, you can record customer patterns on your website and in your stores.

Oftentimes, security cameras are only used to ensure that someone is caught on camera in the case of criminal action. However, if you use a kernel algorithm that determines populated congregation, you can find areas of the store that are rarely visited by individuals. You can also find areas that are highly visited by individuals and then you can make an associated decision to bring the lesser seen parts of the store towards the parts of the store that are popular, which increases the likelihood that those products will be sold.

While you can perform this artificial algorithm on a single video, a programmer would not be able to create renditions of a 24-hour format that showed you the locations of everyone in the store if you had more than 6 cameras. It is a mathematical monument a single person just can't climb and is difficult for even a team. However, an artificial algorithm and an artificial intelligence would be able to watch and make recordings of all video at the same time and produce those results onto a map of the store as it is covered by the cameras. This allows for an extremely accurate targeting of redistributing products for sale.

Website Security Holes

Website security is not really taken seriously until you get to about 120,000 customers. This is usually the range at which hackers become a little bit more involved in trying to get the information out of your database. That isn't to say that there won't be attacks before that, but usually, those are bot networks trying to get admin access or hackers that just happened to find your website.

However, due to the common practices of a hacker, you can use artificial intelligence to find security holes in your website. As I already mentioned, at the lower levels, bot networks will try to access your admin page and it will usually assume you are a WordPress website because that is the most common type of website there is. Understanding this allows a neural network developer to implement an artificial intelligence that looks for common security holes in your existing website and website components you plan to roll out later on.

Less Used Features

Another component that artificial intelligence can handle is determining which parts of your website are not used or which parts of your product is not used. Again, these are normally handled by heatmap with front-end developers and designers as well as marketers utilizing things like heat maps. However, the problem with utilizing heatmaps is that you have to check this on a constant basis as to whether your website is still being used the way you want it to be used.

This means that you have to keep someone hired that can understand how this component of marketing works and, essentially, waste time on a repetitive task. Instead, to save money, you can develop an artificial intelligence that will keep track of all the heat maps since its creation and beyond to determine what customers will be more likely to use in your product than what you have now. Not only will it determine where you should place things on the website, but if you have a program like Photoshop then there might be a classification of tools because of their location and the difficulty in understanding them. Artificial intelligence would be able to keep track of every component on both the website and in the product and provide you with reports on that information without you having to pay someone to do it by hand, which would usually take much longer.

Massive Monitoring

Production Output

As previously mentioned, artificial intelligence is really good at noticing oddities inside of your work line. Provided that you feed artificial intelligence the right information, you can actually find slow parts of a production line and determine which problems that production line part is experiencing that causes it to be slow.

Part of ensuring that you can keep as much output as high as possible is also ensuring that you are constantly removing the parts that are slow. Artificial intelligence can keep track of the entire system and directly tell you what is slow and even can suggest how to fix it. It's a lot faster than human conducted evaluations and a lot less error-prone.

Worker Health

You've likely seen someone in the minimum wage industry that should be home because they're sick but they're not at home because they can't afford it. Worker health is caused by a few factors. You've got possible contamination in the air, possible contamination from worker to worker, possible contamination between customer to worker, and the possible contamination of the product. The only part that you cannot measure is when the worker leaves to go home.

By having air quality sensors that can detect what's in the air by volume and infrared sensors as well as microscopic sensors to detect what's on surfaces, one can utilize artificial intelligence to ensure a company's area is less likely to make a worker sick. A worker being sick means that a customer could have a bad experience, a product could leave the warehouse in a lot and be infected and grow in the environment that it is put in, or the company has to suffer monetary loss as a result of sending the worker home on sick leave. By ensuring the quality of the company area, most of these issues can be avoided.

Sales Versus Market Trends

There are certain market trends that directly affect the sales of a company. For instance, when Apple releases a new iPhone, there is a high possibility that the previous generations of iPhone will all get cheaper. If you sell a competitor phone that competes with one of those older generations of iPhone, you will undoubtedly feel this in your sales when that iPhone is released.

This is just a simple example, but you can usually tell that a product is about to release because the company will receive a huge increase in their market just before they release so that everyone can cash in on the sales of that new product before they cash out. Noticing these trends is quite difficult because the stock market has over 500 companies. However, an artificial intelligence can make short work of such an issue and would be able to notify the sales team that something might occur given a certain company's market price. This allows you to develop a plan of action before your sales see a hit.

Customer Service Speed

Artificial intelligence is really good at handling basic necessities of everyday human life. Customer service is what served this purpose before this purpose got handed over to artificial intelligence. An incredible amount of time is wasted at customer service on redirecting the user to the part of the website where the product that they want currently resides. An enormous amount of time is wasted on customers that are asking simple questions such as "where is the pricing" and "where are your other products". If you have a website that is not particularly navigation friendly, customer service is usually the one that steps in to help customers navigate the website until the problem is fixed.

Artificial intelligence can take their place based off of what they use as answers and, in addition to that, artificial intelligence can notify you that there is a problem concerning a specific part of a specific page.

By handling basic issues, artificial intelligence allows customer service not to be bogged down by rudimentary tasks and allows them to take care of the more agitated customers that are dealing with more complex issues.

Customer Acquisition Rates

By incorporating these measures, one can ensure that products are delivered on time, products match the quality expectation of the customer, and that problems are solved on a timely basis according to customer expectation. If another customer experiences a better product with a different company, they are likely to leave your company as a result. If another customer likes your products but tries not to buy them on a regular basis because they don't like how long customer service takes, changing how you handle customer service would revolutionize those customers.

All of this boils down to being able to keep customers longer, make more money off of them, and increase the amount of money you make overall by optimizing every part of your company.

Micro Financing

Risks for Startups

Deep ROI Analysis

The primary contention point for many startups is determining whether an idea is really worth it or not. The primary way that startups figure out whether an idea is worth it or not is usually by running a return on investment analysis. The problem is that a return-on-investment analysis can take quite some time.

The reason why it takes time to do a return on investment analysis forecast is due to the primary issue of how long it takes to gather the necessary information to do the analysis. You might be thinking that the analysis would be based off of company information, which is correct and rather basic to acquire, but it also requires data from at least five years in the past. Data acquisition and running analysis on 5 years of data for x amount of companies is a mathematical nightmare. If you have 20 companies that would compete with your idea, you essentially have to look at and understand 100 years of collective worth in these other companies just to see if your idea is worth expanding on.

Artificial intelligence can be set up to not only acquire the necessary data but also to analyze the data so that it can be visualized. A startup company is usually interested in how much money can be generated within the first year, how much money can be generated in the years after that, and how much it's going to cost to make that money. Artificial intelligence can not only gather the necessary data, analyze that data and visualize it, but can show you outlier situations that you might have previously missed. Using clustering algorithms, one can associate what profitable practices were put into place during what times a company had been doing before and after the release of a product. This helps the start-up to not only gauge whether the ideas are worth it but also understand the past mistakes and successes of other companies.

Audience Analysis

Furthermore, startups can get key insight into what type of audience they should really be marketing to. It has been a long time since a brand-new company doing a brand-new thing came onto the market. There are new technologies that are coming out all the time, but the reality of it is that the different genres of companies have already existed for at least 30 years. Therefore, those that buy into technology really have the categories of only buying into conservative technologies, mixed technologies, and risky technologies. You might think that Facebook is different than the Oculus, which you would be correct in its utilization, but they would be categorized in one of these three categories. Oculus would be seen as a risky technology as it requires a lot of upfront investment on the consumer behalf to just use the technology whereas Facebook is a conservative technology because it barely requires anything to work beyond a working internet connection. The combination of the two make a mixed investment, so you may only want to use specific parts of a company.

The reason why this is important is because you can create these categories for the area that your startup is in and run an analysis of competitive companies to find out if more investment into a particular part of these categories made those companies more money. For instance, when you first set up an analysis on Facebook properties, you will likely set the Facebook company itself as a safe investment, the Instagram property as a mixed investment, and the Oculus as a risk investment. Then you find out how much money was invested into each program or how much money each program costs and what that program made in its return on investment at the time it was bought. You can then see whether a company is more invested in taking risk ventures, mixed ventures or safe ventures.

The importance of this comes down to the fact that you need to understand the path your company should take as a startup. If you are going up against your competitors, for the first year you want to (at least) follow the common practices of the successful companies you plan to compete with. You don't exactly want to go into a market showing off a risky investment when all of your competitors made their successes primarily off of safe investments. Safe investments allow your audience to check out what you're offering before they determine whether a risky investment is worth it or not.

Therefore, if a company invests more in the safe investments market then you understand that they are primarily about ensuring utilization amongst their tools. If a company is more interested in risky investments, then you understand that that company is more interested in risky investments and makes more money by creating new products to lure new customers in. If you are attracted to the prospect of new customers, then you might want to have your startup be a risky investment but if you want a slice of the pie of the audience that goes with the safe investment, you might want to make a product that adds functionality to an existing product to make it better than your competitor's product.

Future Working Capital Analysis

Understanding the potential return on investment that you might get given a specific idea and matching it against competitors and then beginning to understand the audience of your competitors means that you can begin understanding the future working capital.

Big companies have a huge problem and that is that they change very slowly. The one benefits that a startup has over every other business in the industry that's bigger than a startup is the ability to change quickly. This means that you are used to having better technology, a different game engine that has more features, and you generally have a lack of assets you are liable for. Understanding how much working capital that your competitors annually work with helps you to understand where improvements can be made to your business practices before your company starts. For instance, if your company runs a factory and that factory is brand new but has artificial intelligence in it, you can actually optimize how much your factory outputs versus your competitor who likely did not have the ability to incorporate artificial intelligence at every level of their business and continue operating. The difference between one and the other is that a fully optimized factory is capable of producing more and staying in production longer then the competitor.

By being small, you have the ability to rapidly change without suffering a heavy affect most of the time. This means that you can drastically change your return on investments. For instance, let's say that you want to get into the Farmers market because you think you have a product that would speed up the amount of food farmers produce. If that product is sold in stores, you are likely to be charged a nominal fee for having space in that store. This goes into your overall liabilities and makes sure profits suffer.

Let's also say that the vast majority of small-time farmer product producing companies have either a barely functional website or no website at all. If you can sell these products online versus selling them in the store, you can create an online only business where you are not charged that nominal fee that goes into your liabilities. If a large company tried this, they would have to determine if it was more beneficial to sell products online than it was to sell in the store because they have had success in the store, but the online world is far more competitive. They have to make sure that their change will be profitable beyond what they are currently making whereas you, as a startup company, are able to make that decision because you're not making profits in stores but you do have a product to sell and you have access to an online platform that allows either companies or individuals to buy directly from you.

This example goes even further because most companies will prebake items based off of past orders. They will not do on demand product making because they usually can't afford to as it takes too long in the chain for stores to order new items and for that order to be delivered and processed by the factory. Remember, it was only recently that lot orders have been performed online via email and digital exchange. A lot of factory companies have rudimentary websites that are very difficult to navigate. You, as a start-up, do not have this problem and are able to process digital orders within a timely manner with an automatic system powered by artificial intelligence that can make products on demand. This means you can maximize savings and product sales whereas the bigger companies have to take a hit in production savings because they simply do not have a functional system that allows them to do the same thing as you do.

Risks for Small and Medium Sized Businesses

AI Shows Expansion Areas

When you are a medium sized business, let's say a consignment shop, there are at least 10 different variables that play a role in how much money you can make. This is the primary problem that companies like these have to contend with in order to understand whether they can expand their stores or if it is more profitable to stay as a single store.

The problem here is that most of the issue lies around population density verse is the active, ongoing sales count. Moving to an area or expanding to an area often represents a risk of investment without sales. A normal, small business is not going to have the research team capable of formulating and understanding the best areas to expand to. That isn't to say that they don't have the necessary equations to make a business decision, it's just that there's so much data that requires processing that there isn't enough personal thinking power that allows the business to make the best, most optimized decision as to where that expansion should take place and whether there should be an expansion at all.

Artificial intelligence, on the other hand, can make this process smaller and easier. Artificial intelligence can take the numbers that humans provided and produce the visualized results so that it is more easily understandable on human standards. The possible variables of whether this consignment shop will do well is rooted in population density, competitor density, product variation, better alternatives, and current sales versus rent. If you are looking at a large city, you essentially really have to do this same equation for every mile in that City. A normal human could not do this in a week for a city that is 50 mi in diameter, but an artificial intelligence can be set up to aggregate this information and do it within an hour or a few hours.

AI Allows for Rapid Improvement

Due to the effective ability of being able to understand how to improve products, how to expand audiences, how to choose specific expansion areas, and how to understand your products market better, artificial intelligence is fantastic for rapid advancement.

Companies like to either expand or increase product count. Essentially, companies like to expand on the land or expand in their product line. If artificial intelligence is set up correctly, it can do this on both fronts. Artificial intelligence can improve your products reach into a market and then also increase that products production rate on the land. After all, the goal of a business is to sell as much as possible as wide as possible. Therefore, artificial intelligence allows small to medium businesses to act and expand in a way that would normally have required large research teams that were worth hundreds of thousands of dollars to billions of dollars just to research where you could expand. To this day, large corporations invest massive sums of money into research that allows them to improve and add products to their lineup while also looking for areas to expand in. Artificial intelligence allows startups and medium businesses to make that research happen without the need for ridiculously large research teams.

The most important thing to understand is that artificial intelligence doesn't need any money, it just needs access to data. We live in an era where there is more data available to companies than ever before. Unlike a research team, data is relatively cheap and is even sometimes free given the circumstances in which that data was retrieved. This means it is cheaper than ever to do extensive and expansive market, product, and chain research into a business.

In addition to this, artificial intelligence usually only requires one very well knowledgeable person that works with neural networks and a relatively small investment in a computer with an impressive graphics card. The investment is so small that it is below 1% of what giant companies like Facebook and Google invest in terms of research but, built correctly, can provide just as much useful information as the research teams that Facebook and Google pay for. Startups and medium businesses excel at artificial intelligence utilization because the companies are not already liable to entire sections of the company that are outdated. That singular person can now serve as your research department.

Data Can Be Sold

Perhaps the best part about data is data can be sold to the right buyer on a consistent basis. You have research groups, companies like yourself, startups, and even big business willing to purchase data if it's qualitative and quantitative enough. It's just that you have access to the data and the people looking to buy either only have access to their data or don't have the ability to collect that data.

Let's say that your customer base has somewhere around 100,000 people in it. For most companies, that is a drop in the market bucket because it is so small in comparison to most countries' populations and even some sovereign areas or states populations. However, 100,000 people can be expanded to a million people as a sample experiment. This is commonly done in scientific areas because usually a big enough sample can represent the entire portion within a high accuracy, enough to offset the need to gather expansive data.

The benefit to be a company that can gather data and sell it is that you can charge maybe $6 per person on your platform. That's over half a million dollars annually if you can sell it all and it is just information collected on your users, it doesn't require any additional investment beyond what you would normally invest to benefit yourself. Companies like Google and Facebook do this on a regular basis because it's their primary source of funding. They primarily sell this information to advertising companies and create barriers to accessing all of their data for privacy reasons, of course, but for the more common-sense reason is that they can then the charge more money by sectioning off the data. Knowing this data allows you to understand your user base for the market you are in, thus it is very valuable.

This is simply raw data that's being collected by your server that is already at your fingertips that you can make money off of. This is fantastic for startup companies because it is a selling point for practically all startup companies. Startup companies collect information that is rarely obtainable, which is information that is associated with starting a company rather than a company halfway through its existence. For instance, Instagram was a company that barely had any user statistics in the very beginning but quickly adopted a platform that did so. As it grew, it was able to sell that information to advertisers and begin selling advertising space as they literally jumped leaps and bounds based off of that user data. User data is collected and analyzed with artificial intelligence and if you don't have artificial intelligence steering the information that that data, you're not going to be able to make sense of the sheer volume of data coming from the users of your platform. This may seem like it's an Internet only thing but if you've decided to make a new fridge or a new toaster or anything that can have a chip put into it, that device can send information back to the server and that means that products can be enhanced based on what you sell and how the user uses your product.

AI Result Data Can Be Sold

If you don't like the idea of selling raw user data, there is now a market for selling the results of what you're AI comes up with. You see, companies always try to find a way to lower the cost, an issue dated by certain programs. As I already told you, artificial intelligence usually requires quite a bit of basic investment in this structure of how it's built in the beginning. Having said that, companies don't normally understand how simplistic it is to start with artificial intelligence and so they usually hire separate companies to do the work for them or they just pay the company who are gathering the data to do it for them. This means that you don't have to hand out raw data to buyers and you are able to actually have a better PR presence in saying that individual data is not handed out to third parties.

Facebook and Google have capitalized on this because you don't receive the entire database of information, you just receive the most relevant information for your situation. The most common online tool that is used by most online companies is Google AdWords, which finds the correct search terms that are relative to the topic or product you are trying to get to the customer. This is an excellent keyword search engine that essentially only provides you with keywords associated with topics or products that you want to sell. This saves you on time in researching through the data yourself to find out the numbers for yourself.

They've essentially saved you, as a company, an extra step in the process but they also get to charge you for that additional step while also being able to sell other products in the same basket. These are companies that are specifically built on selling data collected through its products and they only sell the results of that. You, as a start-up, can sell this information to other companies that want to buy it from you without giving out personalized data. This allows them to make their own assumptions based on the AI results and you can avoid scandals such as sharing how much each individual purchase one of your items. You can essentially just do equations for the company that they will just go ahead and do themselves and then charge them for doing that equation. You also don't need a massive user base for that data to be useful, you don't need millions of users and in fact the price only goes up with the number of users that you have. Usually, once you pass 50,000 people you can begin selling data to companies interested in your type of startup. This is due to the fact that after 50,000 people, customers usually see it as a successful business and companies are able to put more trust into that startup.

Risks for Large Businesses

ROA Opportunities

Sometimes running a large business comes with quite a few different costs against your profits. The problem comes in the area of when you are attempting to figure out which areas of your company are costing you the most money.

Often, the easiest way to do this is to determine a return on assets and this allows you to assess each individual asset as individuals or as a group to determine the overall worth they have to your company. The sad thing is that sometimes you can generally do this with employees within the company thanks to performance charts, but business is business. Understanding how much an asset costs versus how much that asset makes you money means that you are able to differentiate between assets that actually cost you money, assets that break even, and the assets that make you money. Needless to say, you sometimes want to get rid of the ones that cost you money or that just don't make you any money, but how do you determine if that is going to harm your business?

For instance, how do you know that a particular asset is not something that a profitable asset relies on? This becomes a lot more complicated and it requires a full understanding of how your facility works.

Artificial intelligence can solve this problem because you just have to determine which components lead to which other components in your profit pipeline. You can then run a ROA and determine whether the asset negatively, positively or doesn't impact the other assets in that line. If that negative asset is vital to other profitable components, which can be determined by your staff and engineers, then the artificial intelligence will calculate that into the return on an asset analysis.

Asset Acquisition Analysis

This set of calculations can be performed on whether you need to acquire new assets and just how beneficial acquiring those new assets will be to your profit pipeline.

Again, unlike other artificial intelligence forms, you essentially have to have your staff or employees that directly understand how that line of machinery or technology works to put in the necessary information to the artificial intelligence so that it can make a quantitative decision on how it will affect your profit pipeline.

Improve Lacking Sections

By having artificial intelligence consistently monitor and regulate machines in your profit pipeline, you can consistently get high profit margins when you run a return on asset analysis. Maximum profit that is a specific asset is giving you, which means that you can easily pick out sections that are low in profits and replace them with new assets that increase profits.

This ultimately means that you are improving the overall system and thereby increasing your asset to profit value, you are actually increasing the quality and production rate of a particular profit line.

Customer Understanding

All of this boils down to maximizing profits and understanding your customer better. Your customer is ultimately what pays for everything in your company, which means that your customer will invest money in working solutions that they prefer.

By being able to monitor everything in your factory, facility, and technology you are able to fully understand every working mechanism that makes you money. By combining this with information gathered from your website, product sales, reviews, and other areas you are able to quickly adapt to changes in the market and apply new products to old product lines. This means that you are able to make money consistently, increasingly, and without much risk involved because it's all based on accurate data.

Chatbots

Getting Rid of Easy Problems

Finding Products or Areas

A lot of time in the customer service area is simply wasted on directing customers towards the area of a website or land mass or towards a product they can't find. Normally, it's usually due to a layout issue of that area that they are looking at that causes this problem.

Artificial intelligence can pick up on keywords like "can you help me find" and then find the product page or area they are trying to find. This is a very simple interaction that takes time from the customer service and that can easily be handled by artificial intelligence. A customer simply logs into a website, starts searching around a website, and then reads about a product that sounds awesome, but they can't find it. That's when they usually reach out to customer service to help them find it. While you are fixing common issues so that it doesn't happen again, you can have artificial intelligence do the redirecting across all your websites and project pages so that the issue is immediately resolved, and this can be logged into a database of queries.

Simple Changes with Billing

In companies like Comcast, electrical, AT&t, and other companies that require a subscription on a monthly basis, a user is likely going to need to change information about their billing arrangements. Simple billing changes would be a change of address where the individual can't find the location to do that themselves or changing the bank account that the account is associated to. These simple changes can be handled by artificial intelligence because they happen in a predictable manner.

Customer service answers the phone and unless there is a specific template set up, a lot of the time the agent winds up wasting is time figuring out the wants of the customer. Once they can find out the wants of the customer, they then spend time figuring out how to do it for the customer themselves. Normally, this interaction takes anywhere from 10 to 20 minutes while putting an additional wait time for the customer at around 10 to 20 minutes. By using artificial intelligence, one can essentially save around half an hour per phone call.

Providing Remedies to Complaints

Another portion of time that is wasted on customer service is customer complaints. Let me use a Comcast example of how to find out whether your internet is currently down or not. If you were to rewind by around 5 years and called Comcast to see if your internet in the area was down, you would call the customer service line and ask why your internet is not working. If you're a little bit tech savvy, you would have done the rudimentary steps beforehand. However, most people simply just say that their internet is not working, and so half an hour of their time would be wasted on diagnosing the issue.

I don't think I have to inform you that waiting an hour to just find out that you have an internet shortage in the area is somewhat frustrating and definitely agitating. Now, if you call Comcast, they have artificial intelligence handling the problem. The artificial intelligence takes your phone number and compares it to the billing arrangements that are attached to that phone number. They then check the address of where you're located to a database that has a listing of current outages as well as repair times. The artificial intelligence then informs you that you have an internet outage in your area and gives you a specific time at which the problem will be fixed. A call that was previously taking an hour per customer was reduced to artificial intelligence providing an answer in as little as 10 minutes from when you first make the call to customer service. When you are dealing with millions of customers, going from 60 minutes to 10 minutes drastically changes the amount of time available.

Noting Easy Answers that Don't Work

As I alluded to before, sometimes simple answers don't work and need to be refined. The problem is usually that the customer is not adequately providing answers to the artificial intelligence and the remedy to such a problem is to refine the question that you're asking. It's as simple as an interview. You will get as good information as you want provided you have questions that specifically target that information.

Therefore, if someone says, "I can't find the about page" and you link them to the about page via artificial intelligence, but it doesn't actually have the mission statement that they were looking for, you have a dissatisfied customer and an algorithm that gave them the wrong answer according to their opinion. At this point, the algorithm should point the user to an actual customer representative to help them and then note that what they were really looking for wasn't on the about page. Then you go into the artificial intelligence and create a condition where the artificial intelligence asks the user if they are looking for the mission statement. Now the artificial intelligence can handle all further customers looking for just the mission statement of your company because the customer usually thinks all mission statements are in the about page.

Personalized Attention

A lot of customer service time is wasted because they are trying to gather information on the customer that they are talking to. They are trying to find what the original problem was and being introduced to a stranger, the customer service representative has no idea what they really want. By having an artificial intelligence begin asking questions, the customer service representative can move past figuring out what the customer wants or, at the very least, they are able to shorten the discussion to an incredibly small length so that they can get to the solution much faster than they would have.

Autobots Defending from Autobots

Chat System Security

A lot of people conceptually do not understand web security because they think of web security in terms of their credentials used to sign in to the website. Web security specialists look at far more than just trying to get to the credentials. They look for any security hole and possible exploits, which some of them use and some of them don't.

The chat system is a dangerous connection because you are directly connecting the customer to someone that may be inside of your business. Usually, a customer of Comcast would be contacting someone with the ability to make account changes in that customer's account. Knowing that the chat system is a direct line from a person of unknown origin to a person that has control over a segment of a company, I think you can begin to see why chat system security has become a prevailing security issue for experts around the world and artificial intelligence is helping in that endeavor.

Predicting Human Behavior

The first thing that artificial intelligence does is try to predict what the other thing on the other end will be doing. Human behavior is, sadly, incredibly well predicted psychology that is easily exploitable and there are a range of job opportunities in exploiting human behavior. Martial arts (not a crime or bad), thievery, scams, Nigerian princes that don't exist, and all the like were all based off of human behavior. Artificial intelligence is able to tell the difference between a bot and a human and the most recognizable to us is in the CAPTCHA that most websites have you click on.

This little box actually predicts human behavior in click patterns to determine whether the person is a human or not. You, as a human, cannot see in infrared so if there is an object in an image that's hidden, that only a computer could see, then it is a very easy way to catch computers trying to get access. You may think that that is a silly way of explaining it, but computers only see binary bits and so if it recognizes the shape of a car in those binary bits then it will recognize it and try to click on it. In addition to this, we have irregular mouse movements as human beings where has computers will take a more linear approach. This is just one way that artificial intelligence tries to prevent other artificial intelligence bots from gaining access to the system. The chatbot system is a window into the company and as such, such artificial intelligence helps to secure that hole.

Customer Provocation Detection

Sometimes the customer is overly irritated at something that currently happened, and they are ready to go ahead and try to start a fight with a customer representative. You might think that the fight will happen, a bad review will be posted, and that will generally be the end of that. The problem is that we live in a society where online image is impacted by everything we do both inside work and outside work. Therefore, if a representative of your company starts a fight with a customer and begins name-calling, this is posted on social media and suddenly that representative is now affecting sales.

You can fire that representative but it's not going to achieve anything and it's not going to prevent incidents like that from happening down the road. Instead, you can use artificial intelligence to begin looking at the language that the customer is using and hand off that customer to someone who is able to control their emotions such as a sales manager or someone higher up the customer will likely ask for it anyway. In such a situation, you have prevented a possibly horrible interaction and you may even be able to retain that customer because artificial intelligence pointed that customer towards someone who could handle that customer.

DDoS Attacks

As previously mentioned, chatbot security is quite the issue on the internet nowadays but there's one aspect of chatbots that most people don't think about. Otherwise known as a distributed denial-of-service attack, a DDoS attack on a chatbot is usually not very well defended against. This is because companies have not caught on that if you cannot DDoS a website, customers will still try to DDoS sections of the website to hamper business.

You can now utilize artificial intelligence to determine whether a DDoS attack is happening and how to shut off that DDOS attack by delaying connections to those addresses. The artificial intelligence could see that connections were coming in vast numbers and then leaving only to find that new connections were being made. Artificial intelligence could react to this situation much faster than a system administrator could and would likely prevent the chatbot system from collapsing from a DDoS attack and may even prevent a website from collapsing if the chat is handled by the same server as the website. After all, the only thing that a DDOS attack achieves is tying up the CPU of a server while actual customers are trying to request access.

Defense Against Reengineering

Sometimes companies don't want to do their own work and so an effective chatbot system is actually somewhat expensive to design and implement. These companies will look at your chat bot that's doing really well and begin plugging it with questions that look a lot like code. Artificial intelligence has been able to recognize code for a long time now and so artificial intelligence can quickly assess whether the user is trying to ask a question or if the user is attempting to break into the source code.

Unlike most companies that just detect the most common code patterns used, some companies have gone about this by asking every possible question they can to gather the responses that your chatbot will use. Such combinations will generally be misspelled questions or different sayings of the same question, which is just them trying to get all the strings that you paid your team to write as responses so that they don't have to spend the time writing themselves. After all, once you create an artificial intelligence system, most of the work is down to tedious response writing and so if they can collect the responses from your chatbot, they don't have to spend time developing it on their chat because you gave them phrases that will work. Artificial intelligence is able to detect whether a user is trying to ask the same question repeatedly and analyze the behavior of customers connecting from that area. This effectively stops reengineering of your chatbot.

Reduce Customer Service Workloads

Get Rid of Basic Problems

As I said, most of customer services problems lie in solving basic, common issues. if a customer is looking in laundry products for a scent ball that they can throw into the laundry basket, but they are not finding it because you put it in the beauty products section, they will likely contact customer service to try and find it.

This wastes an extremely high amount of time on customer service because if one customer is going to ask it then you likely have a hundred more customers that are going to try to ask it and they will try to ask it in 10 different ways. This means customer service agents are just trying to find products for a customer and unless that customer is particularly website or store savvy, they are not likely going to be able to find it without the customer service agent breaking it down into English that elementary students could understand.

That isn't to say that customers are dumb, it's just that sometimes things like accents and the way you say things will interfere with the message that's coming across. Artificial intelligence can be set up to handle general requests for products so if they are looking for a specific type of product then the artificial intelligence is able to provide them with a link to where they can find that product or a description of different places they could find that on your website. This ultimately saves time for both the customer and the customer service agent.

Get Rid of Redirection Issues

Sometimes you have customers that want to do more than what's available on the website itself. For instance, 90% of the reason why a person looks at an about page is because they are looking to do more than the basic customer service and are looking to do business to business service. These are companies that are trying to find out what type of company you are and what type of products you will likely be providing in the future along with what morals you hold your company to.

The problem is that most of these types of individuals do the same thing but in different ways. For instance, one customer might look at an about page and then travel off site to look at LinkedIn pages of employees working there. Another customer might look at the affiliated links to find out where the connections to your company are and how big of a reach you have. In both situations, the customer is trying to figure out who you are as a business and artificial intelligence can pick up on that. These can be handled in various different ways such as flagging them in a database so that you know that this particular customer is doing this or you could start up a chat in artificial intelligence and ask if you if the customer is interested in investing in the business. This really is an opinion-based approach because while one could just be cool knowledge to have, the other could be seen as an intrusive way of business.

Specialized Issues Are Faster

Sometimes websites don't work the way that we want them to and so when a customer tries to reach out to customer service to resolve the issue, they used to have to wait for customer service to be done with all the basic stuff and all the redirecting. Artificial intelligence can find out information the customer wants to provide and usually it can pick up on things like a bug or "this didn't go the way it was supposed to". These would be based off of the words in the sentence such as "I think you have a bug in the website", "something went wrong when I try to do this", or such similar phrasing and this can be redirected to a technical support line. This would allow developers to react to situations more quickly than if the user had gone through a customer service representative and then customer service representative figures out that there was a bug and that the customer was really just wasting the time of the customer service agent simply because the customer can't connect to anyone else.

Reduced Time Spent As a Result

Every phone call that is wasted on basic problems, redirecting, or issuing bug reports is a phone call that costs money. Customer service representatives often work by the hour, which means that you are paying them to solve your customer's problems as fast as possible. By getting rid of the majority of the basic problems and issues that your website has by turning this over to a chatbot system, you alternatively save how much time the customer service agency uses in day-to-day problem-solving. While you are likely not going to reduce the amount of time that the customer service area is open, customers who would come back and tie up the lines the next day because customer service agents were offline can now be handled by artificial intelligence if they have a problem while customer service agents are not available. This saves even more time because customers that have to call the next day will cause customers that are calling that day to call the next day and so the issue becomes a compounded algorithmic catastrophe that whines up wasting a lot of time.

Solving Easy IT Issues

Turn it On and Off

Another issue that chatbot systems that have artificial intelligences in them can do is solve very easy IT problems. If you have an internal chatbot system that an employee can bring up at any time, the chatbot system can then walk them through basic steps of it to save time on the part of the IT staff. That isn't to say that IT staff will simply be replaced, but the IT staff's majority issue is dealing with basic troubleshooting problems.

For instance, one of the most basic troubleshooting problems is to just have a user turn their computer on after having turned their computer off. This resets the memory registers and usually gets rid of some of the most basic problems that the user might have such as a logo sticking on the screen when they're not supposed to. Operating systems, while they are definitely better than they used to be, still have common bugs that are simply solved by turning the machine off and on again. These issues can be handled by a chat system because, odds are, if there is a problem then usually there's an IT article that will help them and this brings us to 3-step solutions.

3-step Solutions

Anyone that does not work in IT usually cannot process what they've done after the third step. It's not really a scientific rule but it is a common experience shared by many in the tech industry. Knowing this, you can solve most basic issues by simply providing 3 step solutions that may or may not work.

If said solution doesn't work, then you can always use the artificial intelligence to submit an IT tech problem into a database that doesn't require the IT staff to be on the phone. The IT staff would then be able to log issues into the database to see if there are common problems or not and find ways to prevent problems that are common.

Reducing IT Confusion

By being able to log issues into a database and have basic issues handled by artificial intelligence, IT confusion can be reduced as a result. Instead of the user calling the IT department and telling them "my email won't work" when, once the IT professional gets there, the IT professional finds out that this phrase meant that the user's computer was not on, the confusion can be cleared up quickly. All they need to be able to do is talk to a company chat system that will begin asking standard, scripted questions that will help narrow down the problem. It takes time for an IT professional to walk to an individual computer in the business and if you multiply that by every user that's currently in that business, you can see how much time is wasted by just traveling to a user's station to figure out the problem for themselves.

More Secure Credential Exchange

Sometimes individuals need elevated access, it should be a rare occurrence but it's actually a somewhat common occurrence in businesses. The only problem is that in the real world, people will leave their phones and computers unlocked in public places so that they can go do something really quick. During that time, the entire world has visible access to your system. If you hand over databases as well as where to find them as a result, obviously your entire backend database just got exposed. However, this is different in an artificially intelligent backed system.

In such a chat system, the user will be in front of their chat system or off the network. You can detect when a signal is no longer inside of a building that would provide the normal security you would want around such credentials. In such a case, before that device disconnects from the network or rather when that network is no longer in access, the chatbot system can then encrypt and obfuscate the information shared on that computer. This will help prevent outside leaks to inside information because humans would not be able to make such a quick decision and so artificial intelligence, which can keep track of hundreds of computers, would be able to send that signal or notice when it was off the network or not used for a given time.

Handling Customer Retention

Immediate Gratification

It's important to handle customer service quickly because the longer that a customer has to wait, the unhappier they are. The unhappier a customer is, the more defensive of their wallet they will be.

Artificial intelligence that can handle basic requests and ensure that they get the solutions to their problems releases something known as immediate gratification. Social media websites take advantage of immediate gratification to create addicts of their users so that that user will continue to come back to that website. It is addictive to watch the amount of likes on a video go up, it is addictive to be recognized by YouTubers by simply paying them a bit of money to say some words. These interactions are addictive. Having a positive customer experience that causes immediate gratification creates a reward for that user to continue engaging in that activity. Therefore, having immediate gratification come from customer service is more likely to result in a sale from a new user.

Most of the reason why products fail is because they fail to provide what the customer needs. The problem with companies is that for a very long time you had to rely on, essentially, common sense to tell you what a company would be able to sell to a customer. Now, with tracking and statistics, you can effectively analyze what areas of the website are failing or areas of a product that are not being used to justify their purpose.

A product's value is determined by how much time they save the customer so if your product saves customers time, but the customer does not think that the amount of time saved justifies the product, the customer will leave. The way that companies currently handle this is they usually provide the customer with tons of tools that can help them save time but the problem with tools is how often and how useful those tools are.

You can open up Photoshop and use Photoshop for years down the line, but you may only use Photoshop for the ability to mix two photos together and that's it. Most people don't know that you can use Photoshop to go through video clips and get a picture-perfect moment or that you can take a photo and change the perspective of the photo based on what's in the photo. These are features that are totally alien to the average user menus and other features, which means they are rarely used but are useful in themselves.

Provide Feature Additions

Knowing the features of your product are failing and getting feedback as to why people might not be using it can actually lead to adding to features so that people use it more often. The more features that your product incurs usage in is equivalent to the amount of investment the customer is willing to spend with you.

The average user does not understand that you can go buy a 12-terabyte hard drive and then proceed to connect that hard drive to the internet so that you can connect to it whenever you want a cloud drive. You can do this and save money from ever needing to have a cloud drive but knowing that your cloud drive has backups and that items can be recovered in the possibility of a storm and that your cloud drive will likely still be online make the investment of having a cloud drive not controlled by you a good investment.

The added features of having redundancy and an always online capability means that a cloud drive offers more than just storage. On top of that, remember that the more that the user uses of your product, the more addicted they become to it. Therefore, you see companies like Microsoft and Google attaching their versions of cloud drive to their regular products. Therefore, you can open up a Google document inside of your cloud drive and you don't ever have to waste the space on your computer for work that you might just store on Google Drive and you can access that work even if you're not at your computer and you just have your phone.

Increased Customer Interaction

As I already said, the more customers interact with you on a positive note or on a note of necessity, the more they are likely to contribute to your wallet. Customers are both a form of income and advertisement. For instance, Evernote is a company that one might support but it's a relatively unknown company because it's primarily for people who take notes. However, given the features of Evernote, it's quite easy to say that if you're doing anything that requires you to keep track of something you can usually use Evernote for it.

As a result, someone who constantly uses it to take pictures of sick notes for their kid so that they can share them with his teachers is likely going to suggest this tool to other individuals because of how easy it is to take and ensure you have a backup of copies. The more this individual uses the service, the more they become addicted to it and they want to share their feelings for that service. They become an advertisement.

Understanding The Audience Better

Understanding what customers a hard time have searching for, what product features they tend to use more, knowing the complaints they have, and having statistics on what parts of the website are more visited helps you understand your audience. Understanding your audience helps you understand their needs and leads to more profitable products that are based on this knowledge. All of company is an entity that recognizes time that can be saved for the customer and provides the solution that saves that time.

Autoresponders

Campaigning like GetResponse and Aweber

Customer Engagement and Lead Selling

Products and services like these allow singular individuals to have just as much impact on customer engagement and selling leads as a full team of marketers might have. Products like these provide commonly used themes so that you don't have to reengineer what other companies have used in the past. You can focus more on engaging the customers when you want to engage them on specific products and you can use this to sell leads.

The easiest way that you can build a list that you can email market to is to provide a product that will essentially fit the needs of the customer while also just being a small treat.

You don't want to give them all of your solution at once, but if you give them a small portion of it then the human predictive habit is that that person is going to want to finish gaining the knowledge or products that you have that might make them successful. This has been achieved by many companies in the past and they only request emails so that they have a list of emails they can mark it to. You do not need to worry about whether your emails will be thrown into the trash because if what you're selling has value, you will have customers out of a percentage of the emails that you collect.

Easy Implementation and Management

When running a marketing team, you likely have to deal with several individuals just to understand where you want to send your campaign. In addition to this, you have to engage programmers to ensure that the email is sent out at correct times on correct days for correct reasons. Essentially, the problem with marketing in emails in the past has been the amount of management overhead that's associated with email marketing. With products like these, you are able to not only easily manage those emails but are able to quickly conceptualize campaigns and Implement than faster than ever before. You don't have to have a marketing team because they have working models for given campaigns that have proven to work in the past, so you don't need design specialists and marketing teams to work on this item if you don't have that capacity. These products essentially reduce the amount of management need on your behalf while getting a similar to even better result because it's using data that has proven to work in the past.

Event and Time-Based Responses

The most interesting concept about these products comes from the fact that you can either have it be setup to instantly respond to customers via email or have time-based responses. To give you an example, if a user wants to look at a certain product and begins to go through the process of inspecting that product, if that user tries to leave you might want to try to sell that product to them at later date down the road. You already know that the customer is interested in that specific product, what you don't know is the reason as to why that product was not something they bought at the time. The most effective way to engage a consumer on a product is to find out why they didn't buy it in the first place, which can be done through email marketing by asking the direct question to the customer. If the customer leaves that product page after spending a significant amount of time on it, odds are they will have a reason why they ultimately didn't buy it. It is far more useful to collect that information because if one customer does it you are likely experiencing it with hundreds of customers.

In another aspect, you might want to have time-based responses. Therefore, if Black Friday or their birthday is coming up then you might want to offer special engagement prices to lure them into your website. This might be utilizing information about their past purchase history through their account on your website to determine the most common products they are most likely to be interested in. For instance, you might suggest that they treat themselves to a high-end headphone set because they are primarily interested in PC components related to gaming or production.

Getting Personal with ManyChat

Easy Messenger Bots and On Demand Responses

Messenger is used more often than email nowadays by those who are willing to spend money more. If you have young adults or teenagers that are in your market, they are far more likely to spend more money because they don't have the financial conservatism that their older counterparts might have in the future.

Messenger Bots allow instantaneous responses that users willingly sign up to receive marketing messages. It is such a completely bizarre concept to understand but if you sell products that a specific type of customer likes, they will willingly provide access to their messenger so that you can advertise to them from your messenger.

The best part about messenger Bots and artificial intelligence that able to engage through the messenger is the fact that it allows you to have on demand responses as users sign up.

Personalized Access Unlike Email Responses

On demand access means that you have personal access to individual customers immediately. Therefore, if you have common questions that are asked by customers you can assign the artificial intelligence to have automatic responses to those questions. However, sometimes customers have questions that cannot be automatically answered by artificial intelligence because of how uncommon they are. Let us say that you sell educational resources that allow users to learn what you have learned over time. What happens if a customer of those resources has gone through all of them and still wants more? The only option is to provide additional resources or to provide personal mentorship, another set of products that you can provide.

If a consumer wants to learn even more from you but has already gone through all of your stuff, you can provide additional resources. The problem with being able to provide additional resources is that it shows where you can improve your website or service so that those additional resources are no longer causing customers to leave your website. On the hand of mentorship, this is an exclusive service and depending on how many people want to be mentored by you, you can charge ridiculous rates for the amount of time spent on the individual. Either way, both of these require you to personally respond. The benefit of having a bot messenger is that the customer doesn't go ignored and you are able to be there and provide them with a solution, a solution that costs money or shows you how you can make more money.

Broadcasting Like Email

The last primary benefit of a messenger bot is that you can still run practices like you would in an email. For instance, a key feature of email marketing is how many members of your audience you can target with a broadcast message. A broadcast message in email marketing is where the marketing is no longer targeted to the individual but to everyone in the email list. You can still do this with a bot messenger without giving up the personalized access that a messenger bot gives you.

More Precise Targeting

Important Personal Days

With these tools, important personal days become exploitable by the companies that have access to them. Let's say that there is a family reunion going on and you are a catering service. In the past, you would not have known that there was a family reunion going on but thanks to access to social media and profiles, individuals share this information on the internet freely. On a regular basis, you can target personal days for individuals to create engaging customers that would normally not have had access to discounts on your services because you generalized the discounts too much.

Serving Sales on Previously Viewed Items

Unlike a store where everything is on a shelf and the success of a product is viewed by how much of that product is sold, you can increase sales on failing products if you know it has interest. On a website, you can tell with heat maps and visitation counts whether a product is being viewed at a high rate or low rate and whether that product is being sold or not. Unlike the model where you simply have a store that has shelves that allow them to bring products to the customer, a website is able to discern whether the product is bad or there is something hidden in the product that's making it bad.

Let's say that you are trying to sell a tactical survival knife. This tactical survival knife has the ability to act as a rope, a knife, and a flint stick as needed. You might not know why the product is not selling and in a store it would just eventually be taken off the shelf. However, on the website, you can track where users are going on that product page. The user will often stop reading just after the part they have a problem with. If you were to go back and look at the product page that had this tactical survival knife, you might see a trend of users leaving once you tell them the material that the rope is made of. It may be a faulty type of rope and you may need to redo that part of the product. You can further engage the customer by asking why they did not prefer that tactical survival knife and get a more direct reason.

With the first option, you generally know what's causing the problem but, with the section where you get a direct answer, you can confirm that knowledge even if the response count is exceedingly low.

This allows you to modify the product itself for the purposes of future sales or to provide that knife at a discount and ask the customer if they're still willing to buy the product but at a discounted price. This allows the customer to choose whether it's too expensive as a product or if it's a fundamental problem in the property, which ultimately chooses which path you are going to go down when making sure that this product is profitable in the future.

Give Personalized Freebies as Lures

If a user is likely to be purchasing survival gear, they are going to want to know as much information as possible as it relates to survival. If the user tends to purchase a lot of headphones, they want to know how they can get the best experience out of their headphones as possible. It takes almost nothing to make an e-book nowadays compared to when writing was not electronic. You can go on websites like Upwork and hire a freelancer to write your book for $50 and then be able to use that as a marketing lure to bring websites into your website to buy your products. These freebies essentially allow you to lure customers into your website much like you would lure an animal out in the wild. The investment costs next to nothing in the grand scheme of a business, it's easy to produce, it's personalized, and it generates website traffic that will lead to sales.

Reminders for Regularly Ordered Items

Amazon released a perfectly designed device that is almost comical in how evil it is. However, they set up a very noticeable trend by doing this. If you sell something like matches or educational material or templates, these will usually be bought on a regular basis by your average consumer. The device that Amazon released was a button that could be attached to magnet capable surface that allowed you to order products that you had already ordered on a regular basis in the past. They were so much of a 'genius' with this that other companies began the to involve themselves into this practice. There are now over 180 Amazon Dash buttons in existence and of those you will find companies like Mucinex, Tide, MegaRed, Dasani and Quaker Oats.

Reminding customers of items that they regularly order invites them to visit your website. You can do that with these products.

Continued Activity and Shared Selling

Follow-Up Information on Offers

A common trend in online websites is to spread out the effective sales of Black Friday. The reason why these companies try to spread the deals out over a given set of time is to allow people to partake in these deals more than they would regularly. The problem with store locations is that they can only handle so many customers at once because not only is it unsafe to have a massive amount of people in one crowded area, but the stores are only able to hold so much product. In the online world, these two problems don't exist for those companies that are participating in Black Friday. They can handle a practically unlimited level of customers and they have almost no shortage on the products themselves beyond soft limits, where they have the product, but they only want to sell so many of that product at a discounted rate.

You may have special details and offers related to the services on your own website, but sometimes customers avoid these details and offers because they simply don't have the money at that time. You can continue a discounted sale past a certain date by offering non-responsive customers an option to pay for an item later that they order now or provide them with a "We missed you" deal. Pay-later is a practice that is often done with software because software has a soft limit that not everyone can overcome. When a user signs up to a website for a specific trial account, they will likely want to know how they can cancel their account should they not like the services their trial account provides them with. This can be handled by these services, but should they go through the process you suggest in the email they can then be prompted to consider a special deal if they stay. This is more likely to ensure the customer at least pays for the first month of a service based on that deal.

Reminders for Forgotten Items

Let's say that you are a software company that sells stand-alone software, software that you can download and then just forget because it's not attached to any subscription of any type. The problem with these customers is that there isn't enough engagement after the software is purchased and it's usually because they've forgotten the software. The best way to get customers to regularly engage with the software that you are selling is to provide educational material or resource material that is useful for that software that will benefit them. This is known as reminding customers of forgotten items and it increases the level of engagement that the customer has with your product.

Associate Product Selling

The last benefit to having continued active selling points is that if you sell one product and you sell another product that is likely to be used in tandem with the first product, you can more easily sell that second product. Customers usually navigate to a website to buy a singular type of product but if you are able to sell that product, they might be interested in other products that are associated with the use of that product. These products allow you to instantly respond to customers buying the first product so that they understand there is another product that benefits the workflow usually associated with the first product, something that stores don't have an advantage with. In fact, stores usually have to place the items directly next to each other to even try to get the same effect. The most common relatable reference to this is where Amazon attempts to show you related items or items that are normally package together just below the product listing page.

Talent Acquisition / Human Resources

The Wider Reach

More People Faster

A human has to sit down with a list of names and a list of people before they can individually create emails for each person. No recruiter worth their salt will send out an email with every one of their correspondence attached to that email. It is simply Ludacris and it's a bad sign for the people who view that email. Therefore, artificial intelligence allows for a recruiter to reach more people at the same time.

If you think about it, this is a mathematical problem. If you have one email for every one customer for everyone website, then if you have 100 emails for 100 customers over 100 websites you don't have the ability to send that many emails out in a day. However, a machine can do it in a few minutes and a machine can actually do much more than that even though you're able to cast a wider net.

All Platforms At Once

Artificial intelligence can gather all the places where you would normally search for valid people looking for a job. They can filter out the scamming websites looking to get people's money for jobs that will never be fulfilled because no one actually signs up to their website. In addition to this, artificial intelligence is capable of sending out emails to everyone on a specific platform because it follows rules.

If you look at the way that you post a job on Dice, you will find that it is very different than if you post a job on Upwork or LinkedIn. However, there is a series of actions you have to go through every time you post a job. Additionally, every time that someone signed up to your website, you have to handle that job posting through email. Unlike the platforms where people go to look for jobs, people who receive recruiter emails usually accept one small format and has one platform. However, there are different formatting tools for Yahoo than there are on Google. Artificial intelligence can handle all of that and all you need to do is focus on the job description and making that job description personalized.

By being able to reach out to more people at a faster rate, you are much more likely to reach the individuals who would normally have had an interview set up for a different job by someone who managed to reach them first. Instead of having to wait through each and every email, you can instead skip through the wait of sending emails and start receiving applications that day.

However, one problem to being a tech recruiter or a recruiter of some kind that works by email is that the email addresses that you make might fall out of favor, might be controlled by bots, and might actually just be controlled by other recruiters.

Understanding that these people exist, you will either usually get an email back applying for the job in several different ways, replying back with malicious code, or no reply at all. Normally, application emails are so long in how much time it takes to go through them that it is difficult to manage all the waste that occurs when you blast that email out. Artificial intelligence on the other hand is able to filter through emails that try to game the system by filtering out recruitment offer emails, filter through any of the emails that might contain executable code and will mark the applicants who did not apply to the email. That last one is more useful than just ensuring that you don't email that person, because that person may actually have useful skills and you can compare the previous emails that user ignored so that you can filter out the skills that that applicant doesn't apply to so that the job applications are more personalized for them. All of this reduces failure rate at getting applicants for jobs.

Language Translation

Some of the applicants don't speak English but because of artificial intelligence, that is perfectly fine. If your client that you are trying to get a job for cannot speak English and they are part of the group of emails that you send out, you can have them submit a preferred language for you to use. This would be whenever they originally signed up for tech recruiter emails or other recruitment positions and artificial intelligence will take the words that you use and translate them so that the applicant can fully understand what you are trying to convey.

This is particularly useful if the position is willing to relocate the skilled individual as it translates the information into their native tongue so that they don't have to use Google translate or another translator to just understand your email, you've gone out of your way to do that step for them. While we all know that such a translation is not perfect, we also know that the translation generally gets the idea across to the applicants.

In-House Recruiting

Possibly the best benefit is the ability to recognize talent within the company because talent within the company has several benefits that come with it. You already know their work ethic, the price they're willing to go up to, and the amount of trust you can give that individual. An artificial intelligence can check the currently applied applicants to other job listings to see what that applicant is applying for. For instance, on LinkedIn they may list themselves as currently working but they also might have a tagline that says they are currently looking for new opportunities. Artificial intelligence can go through your colleague's social media outlets and find out if one of them is seeking a job opportunity that fills a job area you want to be filled. This saves on time, money, and usually training that comes with new employees that don't know how the company works.

Specific Skill Targeting

Likelihood of Associated Skills

For normal humans, understanding the skills relative to an industry is vital to understanding what you need to hire for your company. Therefore, if you try to put out a job posting with barely any understanding of the skills required in that job posting, you are not likely going to be very effective when it comes to understanding which applicants should and should not be hired.

There is a term in the programming industry called the magical unicorn and this refers to human resource managers attempting to hire programmers at that behest of IT managers that don't effectively explain what needs to be hired. A human resources manager doesn't normally need to know what programming works in what location, they just need to know the type of person that needs to be hired. The problem comes into play when the IT manager doesn't effectively communicate that they may be looking for more than one person.

For instance, a common post that you might see is the requirement that the programmer has at least 8 years in Java development but at least 4 years in Python development. The reason why the two of these clash is that the eight years in Java development means you are generally looking for a senior Java developer but the python requirement of four years usually means you're looking for a junior developer in Python. In such a job posting, it becomes quite obvious from even further requirements that the job posting is actually for two different individuals. For instance, some job postings will say that you need experience with Springworks but then go on to say that you need experience with the math.lib library of Python. Java and Python both have extensive mathematical libraries and to need one over the other only happens in a few special cases. It is rare that you need someone as proficient in Python mathematics as you do in Java web development, but a human resources manager would not understand what the difference between the two are and why they are so separated. The magical unicorn refers to a position that no singular person can fulfill unless they had very specific training.

On the other hand, you don't need to if you develop an effective artificial intelligence that can handle associated skills. On an annual basis, companies run statistics to determine what skills likely candidates will have so a Java developer will likely be more experienced in web development on the backend and a Python developer will likely be more experienced with neural network engineering. Having an associated skills algorithm, a resource manager will likely stumble across the fact that the request for such skills is going to hit a very low margin of people. Therefore, the human resources manager will likely request more information from the IT manager and the situation can be clarified from there. This improves the hiring rate of worthy candidates to positions rather than simply not getting any or getting very low turnout for job postings. After all, the best job postings attract the largest pool of people so that the most skilled can come up on top.

Pick up Young Workers with Social Media

A huge portion of the job market today dealing with business and technology directly revolves around social media presence. Social media is a great form of looking into an individual's personality to determine their effective worth inside of a company that needs an individual to incorporate themselves within that company.

The only problem is that unless you have tons of human resource researchers dedicated to the simple task of looking at everyone's profile, it becomes difficult to utilize this technology to your advantage. You could hire a social media manager so that your company looks good, but talent could go missed simply because of a lack of reach and a lack of understanding of whether those applicants are worth joining your company or not.

Artificial intelligence can work in a few different ways when it comes to social media. For instance, let's say that you are trying to find anyone that is a new developer looking to get a job. You could build an artificial intelligence that follows a few well-known organizations for programming development and then begin iterating through the tweets and replies of those organizations to begin targeting programmers. You can then begin adding those programmers to this artificial intelligence and follow them on that platform so that you can further branch out by looking at those profiles' individual tweets and replies. While this network is expanding, each applicant is put into a search query for LinkedIn or Monster job or Uber jobs, really any job application website dealing with programmers, where it then begins to assess the skills of that individual. You can then begin formulating who is available on the market and what type of jobs the market is looking for, which is a benefit in two different ways. One, it extends your direct reach with currently known programmers and it also allows you to assess what new technologies should be incorporated into your company because useful tech usually has a trend. For instance, Ruby developers in the United States are actually kind of rare but there is a trend that Ruby is being used in massive applications meant to handle millions of users. Understanding this allows you to invest in possible solutions that might make this programming language useful in your current business practices.

Find Talent as They Leave Companies

In that same social network, it is highly likely that you're going to be able to keep track of when people leave companies. Common search terms are utilized whenever an employee is excited or even sad about leaving a company, which they usually share this information to the online world.

This allows you to keep track of top talent that have left industries and quickly monopolize on the situation by grabbing up that talent. Instead of hearing in the newspaper that some other big company managed to grab them, you get notified the same day and usually the same hour that that person has been let go and your company can now acquire that talent. This brings your company positive media as high-profile talents usually get some news coverage whenever they are let go from a company and when they are hired by a new company. For instance, someone that is fired from Intel that goes to work for a microprocessor company might get a lot of news coverage from tech outlets due to a relation between Intel as a company and the microprocessor company.

Assess Separation of Skill Necessity

Sometimes, skills are separated because of bias. For instance, for some odd reason there is a common notion that anyone with programming experience can somehow fix hardware. The reason why this is a common notion is because the average person cannot divorce the concept that software is not the same as hardware. It only requires a small bit of understanding to know that hardware works fundamentally different than software and that the two cooperate with each other.

There are other skills that clump these two areas together where they usually need to be separated. You need a hardware technician to fix electronics around the building, but you need a programmer to fix websites and software applications around the building. They are two different positions that are commonly confused as one entity. You have

113

this same collage of concept when you go to any highly-skilled profession that is out of reach for the average public. For instance, there is a massive majority of the public that doesn't understand that a nutritionist is different than a general doctor. In the same swing, you have a massive portion of the population that doesn't understand the difference between a 3D artist and a 3D modeler. It's nuance in that you need to understand that industry and the components in that industry on, at least, the most basic level to see the difference between job positions. This is essential for hiring the correct candidates for the correct jobs in your company. There is often a high turnover rate for positions simply because the company doesn't understand the position they are hiring for is not the position that candidate is applying for because the job posting is covering as a different position. Due to the fact that artificial intelligence already works by separating values and categories, artificial intelligence can bridge the gap between the lack of knowledge about an industry by hiring managers and where the hiring managers can find the people needed for their problems.

Advertisement Versus Recruitment

No More Text Only

A lot of old recruiters and hiring managers don't understand the importance of a nice graphic to lure potential applicants. Essentially, the old way of applying to a job is by utilizing a demands script that simply list the demands of the company and then the potential benefits that the company will provide the user is starting to not work.

The problem is that the world doesn't generally work like this anymore because the fundamental difference of job necessity has changed. It is now more lucrative to advertise to existing skill what they would get by joining your company than it is to actually try to find were the candidates are. There is a huge gap between those that are skilled and those trying to get into the market, but what hiring managers have missed is that neither of them care about the requirements. The requirements had become a secondary problem and so it is usually only by listing the benefits for applying to a position that lures in the talent needed for that position.

If you are a Java developer who has had eight years of experience, you are not likely jobless. Companies that are looking for developers with more than 2 years of experience assume that the tech industry works like any other industry. The problem with this thinking is it doesn't take into account the huge number of openings for tech positions and these tech companies as their competitors. These openings are usually filled within the first week of application because of how many tech developers usually apply to the posting, which is anywhere between 5 to 10. It's actually a quick process because there aren't many of those excessively experienced developers on the market.

Instead, such an industry really needs to focus on what the company can provide to them if they are willing to learn more or intern at the company. There are far more fresh developers on the market than there are experienced ones, but companies are searching for experienced developers more than they are fresh developers, which causes a huge divide in the amount of tech jobs that are in the market because those positions are not likely going to be filled anytime soon. There is simply not enough people to apply to an experienced position because of how hard it is to get in a position in the first place when you were a fresh developer.

Therefore, companies are now starting to focus on providing fresh developers with benefits knowing they will likely have to train those developers to meet their needs. It is a mutual benefit for companies to do this, but they are putting job postings up like the older times where the company simply listed what they needed and looking for somebody to fill that role.

The reason why understanding this is important is because artificial intelligence can tell you why a job market filled with experienced position openings continues to thrive even though there are so many new developers on the market. Artificial intelligence can analyze the market and come to understand that there is a skills gap that companies might have to assist developers in order to overcome that skills gap. In accommodation to overcoming that skills gap, a candidate will work for less money and provide more investment in the company as time goes on as they are grateful for the opportunity.

Training When There Are No Applicants

The reason why training is needed in the technology jobs market is because there are so many fresh developers coming out with computer science degrees or boot camp certificates that simply don't understand how the corporate development environment works.

Unlike accounting, which is a standard and common practice, technology is like being a new mechanic. You cannot take a mechanic who understands cars from the 1980s and prior and try to have them work on brand new vehicles because there's so much more that they need to know that there is an effective skills gap. Training is needed in such a situation so that the educational system can give companies a baste standard from which they can train students up to the level they need them to be at. The problem is that companies, who have generally not needed to rely on technology, don't understand the common practices when dealing with developers and artificial intelligence and that means they usually cannot judge whether candidates will need further training or not based on their skill sets versus more experience developers. Artificial intelligence can highlight areas in which candidates will likely need training before they ever come into an interview.

Pre-Hiring

This gives you access to what is known as pre-hiring processes. It is a step in the stage where you are able to see what skills your candidates have that you can exploit for your company's business. For instance, usually a developer isn't just a developer. They worked in a different industry for a long time before they decided to switch to a different industry. There are content writers, marketing specialists, and many more that have just decided to switch careers and are vying for positions that would normally require training. This is because they think of themselves as brand new developers when, in fact, they have a lot of the skills that would have been trained in new developers who had not experienced a corporate environment. You get to see additional skills that you don't have to train for or compensate for when you hire that individual, a list of additional benefits as it were. The problem is that unless the individual fits the job description, hiring managers usually throw it away for applicants that are more skill oriented and require less training. This further increases the skills gap in that particular industry and artificial intelligence can inform you of when a particular candidate is less fit for the position, but the additional skills of that individual make up for what makes them less fit for the position. Candidates who are less fit are likely going to be thrown out by hiring managers that don't understand this importance. This is happening in the tech development market, the trades market, and the medical market.

Testing Applicants Before There's an Opening

Once artificial intelligence is able to gather the individuals from the social network, look through their profiles for additional benefits, it's time to test them on what you need them to know. The most common application for programmers is the whiteboard interview and this is an interview where the programmers code is tested against a random interview question. This is something that can be automated by artificial intelligence and the artificial intelligence can use metrics to determine how much value a specific candidate has in comparison to the values of the candidates that are applying next to them. For instance, they may take a slightly longer time to come up with code, but that code is a hundred times faster than any of the code that anyone else has produced. There may be an additional programmer that does come up with the correct code that is slower than the faster individual, but they accounted for memory management and cycles draw. These are both highly experienced individuals but on different levels. If you can't see why one developer is better than the other, it's time to use artificial intelligence. A person who creates an algorithm that is conscious of memory management and cycles draw has written server programs before. A person who writes code that is faster than anyone means they are likely writing code from a previous challenge that was similar to this interview. It becomes apparent that one is better than the other only after the experience of a senior developer is shared with the one that doesn't see the difference. This can be quantified and easily understood if the metrics are gauged by artificial intelligence.

Answering Candidate Questions

Candidates will Interview You Too

Candidates are more aware of companies and their working environments than they have ever been. In the old days of working, prior to the internet and even a little bit into the age of the internet, workers simply wanted a job. However, knowing that a certain skill set is needed and there are only so many positions open creates a problem for the company because now the job worker is looking for additional benefits such as having their lunch taken care of for them, having access to daycare, and similar things. These candidates know that their skills are worth hiring but that companies are only able to compensate them so much, so they are looking for useful benefits that come into play while they are working.

Understanding this, you begin to realize that the candidate will likely be having some questions when it comes to the type of employment offered to them. This means that you will need to answer these questions and if you are a hiring manager, you are not likely to be used to having the candidate interview you and you representing the company in your answers.

This situation can actually be resolved with a chatbot. Normally, the benefits are pretty customary but common. They want educational time that's paid so that they continue to stay relevant, child care because they may be a single parent or both parents in the household work, and maybe they want a retirement fund. These are common benefits and you can actually see them being common on other job postings. Understanding this, you can have a chat system that will answer basic questions like the benefits that the job comes with so that you, as the interviewer, can focus on simply understanding whether the applicant is worth hiring or not.

Save Time By Removing Confusion

The applicant may actually have some questions dealing with company practices or how the process is going to go. You may simply be coming into an interview just to see what the person is like rather than try to get a technical interview that gauges their skill and worth.

A candidate doesn't normally know this whenever they go into an interview, they just know that they are having an interview. If you have several applicants that need the same questions answered you can set up an interview chatbot in a relatively short amount of time that will handle the questions of the developer before this happens. While I may be talking about developers and programmers, it is really only due to the relative understanding of this industry. Artificial intelligence is like this for highly skilled positions. There are only so many doctors in a country, only so many hardware technicians in a country, only so many programmers in a country and because there is a limited amount, this happens in every highly skilled job because of how many job openings versus talent are available.

Applicants May Fit Other Descriptions

You can also use this time in the chat bot to poke and prod and see if that individual has more than just the skills underneath their belt that you got to see. As I mentioned earlier, the fresh developers are usually experienced individuals who have previously worked in other industries. For instance, if the person is a McDonald's business manager you now have access to somebody who can handle project management, with a little bit of training of course. If the person had been in the medical industry and knows how to use medical devices, this could open a new line of products for you that's associated to what you currently sell. Having artificial intelligence gather extra information on a candidate can happen at once whereas a hiring manager can only handle one person at a time. This saves a ton of time when it comes to hiring individuals and it will often lead to more information than a biased hiring manager will give you.

Prepare Them for Their Interview

Not every interview goes the same way. For instance, it is rumored that Google will have 4 interviews and several challenges before they hire a programmer. This is so that they get a full understanding of that programmer and are able to adequately prepare an environment for that programmer. Meanwhile, a startup company will likely do a single interview and simply hire whoever looks best based off of that interview. The interview practices vary by business and so some of the time, inside of an interview, is made up of answering questions on what the next iteration of hiring practices your company follows that the candidate will be subjected to if they pass that stage. The reason why it's important to let artificial intelligence handle this is because it allows the hiring manager to shrink the number of interviewees that they have to deal with. An applicant that doesn't want to wait three to four months to be hired at a position wants to know whether this job opportunity is worth the time or not, which can be answered without the need of a hiring manager setting up an interview time. Instead, if you have 50 applicants that will take a half an hour each and you use artificial intelligence to knock out the ones that are not willing to wait the amount of time that the interview process takes and you use the same process to knock out the ones that are not going to be able to sustain their financial situation through the process, you might have shrunken this pool of talent down to 25 applicants. Something that would take significantly less time to handle.

Reducing the Selection

Filtering by Skill

The main purpose of incorporating artificial intelligence into a process like recruiting is simply because you want to attain a higher quality result in a shorter amount of time. You can not only select freelancers and individuals on a wide scale, but you can also receive their applications and begin sorting them as soon as they have submitted their applications.

The average user will ultimately use the same words to describe commonly shared skills. For instance, when a Python developer is sharing his or her experience as a Python developer in a quantitative format they will usually have the number of years and the words Python development. Artificial intelligence can go through over 5,000 emails that might be applying to a job and then filter out the ones that have less than the amount of experience you need in just Python development. A huge number of applications have common applicable terms that are often used to describe skills.

Knowing that doctors generally have the same skills, accountants generally have the same skills, and high skilled jobs often describe work experience and skills in the same way allows artificial intelligence to create a filter. A filter allows you to take the more reasonable amount of 500 applicants and reduce it down to 100. It allows you to select the cream of the crop without actually having to go through every single email. If you were to try and go through every single email of 500 people and it took you five minutes to go through each email, that is 2500 minutes or around 41 hours to just filter through the list. Considering that's just a normal work week, that is an insane number of hours spent on just filtering out the applicants that don't fit your needs most of the time. Meanwhile, artificial intelligence can do it in a few minutes. You can go through every single email and based off of the categories you set beforehand, you can actually filter out most of the applicants and reduce the pool size that you might need to interview based on the skills that they applied with. It is actually the same way that hiring managers would use except that hiring managers take a long time because they can only handle one process at a time where is artificial intelligence can go faster. This is because it can handle multiple processes at the same time and go even faster because it can look for keywords rather than read the entire thing.

Filtering by Amount

If you remember correctly, I said that you could set up a chatbot that would essentially ask basic questions with the applicant. One of those questions could be a requested salary amount and if you're only looking to pay a specific amount, you cannot only filter the entire job applicant pool by skill, but you can then filter those into separate categories between the applicants that are in your price range, the applicants out of your price range, and the applicants underneath your price range.

The reason why this is important is because you don't always want to be within your price range because sometimes somebody comes from a cheaper place and expects their skills to be worth less at your place. Therefore, you might have a highly skilled individual that is charging less for their skills than the individuals in the category where your best prices are at. You essentially have the option to save money by hiring a less-than candidate that was only filtered out based off of salary level. Again, I feel the need to remind you that this happens in mere minutes. Even if the hiring manager took 2 minutes to look at every application to check their salary rate, it would take a hiring manager that reviews 100 applicants at least 200 minutes or about 3 some odd hours to complete this work. Meanwhile, the salary free artificial intelligence can do this in mere minutes.

More Likely Candidates to Apply

Understanding all of this means that not only will you be able to filter through applicants faster than you probably ever have without artificial intelligence, but the candidates that do apply to your job posting are higher quality candidates. The reason why most people avoid an application to a job posting is because they are lacking two to three skills off of the requirement list.

This actually reminds me of a job application that I saw once that required somewhere around 2 years of Android application development knowledge and 5 years of Python development knowledge. This is kind of understandable and it might be easy to find somebody but you're not likely going to get many candidates. However, this is where the magical unicorn enters the room. The application wanted the user to have at least 4 years of additional experience in a very specific banking software with very specific tools inside of that banking software that directly related to the business portion of where they were hiring this candidate. Now, I know why they did this. They did this because they either didn't want to or couldn't train an applicant to use this software, but the problem is that they chose a specialization of the specialization of a specialized industry. If you took a pool of 100 people and you cut it down by skill to 20 people and then you cut it down by software to two people and then you cut it down by tools inside of the software and you get maybe one person 20% of the time you understand why this is a magical unicorn. It is incredibly difficult to find someone that fits that description.

However, if you have more than one job posting where each job posting removes a single key skill, you can then aggregate all of the applications into a singular pool of applicants, note repetitive applicants, and then begin grading by skill level, salary level, and lack of skills. This allows you to not only have a chance at getting a candidate like this but it also opens up the borders so that you get people with a relatively close experience with exactly what you want and while you may not have wanted to train someone, if no one with that exact experience applies then it's better to do it this way so that you can get someone close to the goal rather than get no one at all. This is how artificial intelligence is able to get more likely candidates to apply to your specific job posting.

Filter Through the First Stage

Now, if you know that you're going to get thousands of applications and you've already filtered them down and you now have hundreds of applications, you can take it a step further. Instead of having them sit down for their first interview, you can actually allow them to take a test and cancel each other out.

You might be asking how artificial intelligence is associated with this. A common occurrence for a cheater is how long it takes them to look at a question, read it, and then answer it. A cheater will go on Google and waste about a minute or two trying to quickly look up the answer even if the test is timed. This is because even if they can't get the exact answer, they can usually get a good idea of the answer. Knowing that cheaters answer questions more slowly than experts, you can begin to filter out the people who want to get the right answer and the people who know the right answer. This is done by measuring aggregate data off of a heatmap while the user is taking a test.

That isn't to say that the information is definitive because in order to have it be definitive, you would have needed to place a keylogger on that user's computer to see what they typed and then a microphone listener to determine if they used Google Voice instead. In such a situation, it would obviously be seen as an invasion of privacy.

However, knowing that cheaters will consistently take longer on questions and more novice people will also do the same thing, you can filter out the ones that know the answers immediately from the ones that need the extra time. This essentially allows you to further refine the selection of candidates that you have.

Save Time By Reducing Selection

I've mentioned it a few times, but the key goal here is to reduce the amount of time and increase the accuracy a job posting, or multiple job postings will have. Normally, the hiring process takes weeks to months to even do and most companies spend enormous resource hours into handling such a thing.

As I have explained, artificial intelligence doesn't take as long as the average person does by a huge amount to the point where most of the work that would take months to do could be reduced to a few minutes to maybe an hour. If you think about it, we are talking about the savings of multiple tens of thousands of dollars dedicated to the hiring process that is saved by an initial investment in an artificial intelligence sorting algorithm. Such a sorting algorithm could be set up in as little as a week and the savings in comparison to that are tens of thousands of dollars and it's not like the thing will ever become obsolete. Every time you go through the hiring process, you can use this machine for whatever system of hiring you're trying to do. You will still save the same amount of time that you did when you first initially invested in artificial intelligence but also save the money it cost to train it as it is already built.

Business Regulations

Stays Relevant On Current Law

Built-In Law Book

We all know that the human mind is not particularly good at having an exact memory. No human mind on the planet can memorize everything they come across, unless you are talking about the singular special individuals that have something fundamentally different with their brain in comparison with the rest of humanity.

Unlike humanity, computers utilize database systems as their memory and database systems are far more useful and accurate than human based memory. If a log currently exists inside of a database, it will always stay the same unless something catastrophic went wrong.

The best part in such a system is really just how fast this information is at your fingertips. Even if all you did was create a database of laws, it would still be useful for not only your company but any other company trying to succeed in that area. In fact, there are databases specifically sold to companies that only provide the ability to look up laws as it relates to that particular country.

However, that isn't the purpose of this database because if you just needed a quick look up then usually there are resources for that. The purpose of building such a database or having access to such a database is to allow artificial intelligence to access that database when it needs to determine compliance. By having artificial intelligence monitor user activity, work activity, the environment of your facility, all the electronics and machinery, and monitor production output means that it has instantaneous judgment as to whether any one of those breaks compliance.

Normally, the only time that a company knows it is not in compliance is when that company purposely breaks the law. Otherwise, most of the incidences are simply forms of neglect because they don't do personal audits to ensure compliance. They don't do personal audits because the process costs a lot of money and small to medium-sized businesses simply can't do this. Artificial intelligence on the other hand wouldn't cost more than a week worth of work by the neural network engineer and a computer because you already have all of the monitoring setup to keep track of everything, you just need an artificial intelligence that will keep it in check. Instead of having an audit every 6 months because you want to knock off the price tag of having it daily, artificial intelligence can give you a daily audit report on compliance after the initial investment.

Self-Contained

In addition to this, one of the hazards that comes with an audit is that unless the auditor is directly owned by the company, which is often seen as a bad thing, the auditor gets to see everything. As a normal person, you might think being able to see everything is not that bad because you couldn't really do much with anything as you couldn't save it or you couldn't take any of it out of the building. However, technology experts will understand very well the concept of stealing ideas upon generation. Many times, there is a process to create new ideas and it's usually a secret sauce the company doesn't like everyone to know, but an auditor might have access to this supposed secret sauce and if that auditor doesn't care about being in their industry for very long, they might try to sell that secret sauce off to another company.

Artificial intelligence is self-contained, which means that it doesn't take that information anywhere you don't want it to and it can be as secretive as you want it to be. Artificial intelligence simply does what it's told and doesn't do any more, unlike an auditor that might be corrupt and want to sell the information that they gather to another company. Of course, you do have the governmental body that does an audit but if the government is corrupt and rips you off, there's not much you can do about it but there is something you could do about it against other companies. Having artificial intelligence not only means that you can prevent outside auditors from enacting corporate espionage, but it also means it can be transported.

Odds are, you're going to want more than one of these and you can save time by simply building a different database and changing the current algorithm to understand that database instead of building a new artificial intelligence. In fact, depending on that artificial intelligence's hardware, you could actually have a central location where all the data comes in and that Central Hub is responsible for publishing those daily audit reports.

This means security can be localized to one area and you don't need to have direct access to the machine when you're in that other country because it's maintained from a central hub. This makes it far easier to distribute the artificial intelligence network and monetize the information in that network. You can monetize the database of laws that you have so that they pay for themselves and since you are likely to be selling a database per area, you are going to get a lot of customers for each additional area you expand to, which are some of the most profit inducing products you can produce as a company.

Automatic Law Updates

The best part about it is that laws are extremely easy to update because all you have to do is pay attention to the law entry of the local government. The local government will make amendments and changes to the law and these are published to the public so that they can understand how the law has changed. After all, the government kind of fails whenever the citizens can't follow laws that it doesn't know about, so you can actually update this law database in real time as laws are passed or denied.

Monitoring Made Easy

The most difficult part about monitoring is consistency because different sensors will do different things and knowing that different countries have different products, you can understand where different sensors will have different labels.

The massive benefit of having access to a centralized artificial intelligence is that consistency can be achieved. You do not have to buy the devices found in that country to ensure that you get the same monitoring levels as you would in your central hub location, you just have to bring the devices over. The reason why is because if you were to hire a human auditor, you might get one that spoke English, or you might have to hire a translator just so that the auditor can communicate

141

with you. The standards by which auditor do their jobs often differs because of personal views about the regulations in place. This even changes with government auditors where some of them are more lenient than others, but artificial intelligence is consistent, and you don't need a translator once the system is built. In fact, the best way to handle such a system is to not only ensure that you are in compliance with artificial intelligence but have a way of proving you are in compliance with artificial intelligence.

Proving that you are in compliance means that you are able to skip the process of where the auditor spends time looking through the building to inspect what needs to be inspected. By having proof or, sometimes, setting certification standards on what is proof, you can reduce the amount of time that auditors need to physically be there from the government. The reason why I mentioned certification is because that is yet another aspect you can use to monetize this platform because if you're already selling the database, you're selling the custom artificial algorithms, then you might be able to set the standards as a certification so that auditors trust in that certification. This might need some working with the government, but if a certification consistently proves to be true then usually the government or individuals in the government that understand the certification will pay heed to the certification. It's yet another way just to make money while monitoring your facilities.

Suggestive Practices

If you develop your artificial intelligence advanced enough, you can get to a place known as suggestive practices. This is where the artificial intelligence notices that you are out of compliance but because of how advanced the artificial intelligence is, it can suggest either how to get back into compliance or even avoid the need to be in compliance.

If a particular mechanism is part of your company and cannot be removed, then you need the suggestions on how to get back into compliance. If there is a method of optimization that removes a component of the process in which that component has fallen out of compliance, you can then follow the optimized path and no longer need to keep track of that necessary compliance measure.

However, there are very few companies on this planet that have developed their artificial intelligence to this point, but they benefit massively from such an artificial intelligence because not only does it keep them in compliance, but it also saves them money.

Simplifies Tax Codes

Built-In Tax Book

Just as with laws, taxes are usually methodical and changes do not happen rapidly. The best example of a simplified tax service is TurboTax and TurboTax actually implements a form of artificial intelligence into their system.

However, due to the relative knowledge between the tax form being filed and the individual filing that tax form, TurboTax takes a gradual step back the more complicated the process gets. This is because the TurboTax team would need to invest huge sums of time to get this area of their software into all possible combinations.

However, the funny thing about the tax code is that most companies file the same things. If you have learned anything about artificial intelligence up to this point, it's that if something is repetitive the odds of artificial intelligence being able to do it is pretty high.

Tax codes in different countries all tend to use the same parameters all the time and rarely do they change on an individual basis. It is only when the government tries to do tax reform that significant changes happen to the tax system, otherwise it's usually a line or two of the tax pages that have changed to clarify what was previously there. Knowing this, you can set up an artificial intelligence that will take the necessary information and plug it into the specific areas that it needs to be in. However, that doesn't require artificial intelligence you might say and that it just means you are making a simplified form.

Flag Loopholes

What does change with artificial intelligence is the ability to find loopholes in the tax system that were previously gone ignored. Now, you might think that such loopholes require for the artificial intelligence to understand context but, really, they just need to be able to go through combinatorial tax reforms.

It's kind of how really efficient testers of tax documents figure out how much money that you as an individual are going to get on your tax refund. Essentially, you take a test case and you begin going through a single tax preparation that is similar to yours. You then begin to run calculations with every step that you take and then you note the areas in which other forms of taxes would affect that area.

Now, keep in mind that this is usually done by an accountant. This is usually what takes a good accountant a long time to do your taxes because they'll try to get the most out of your taxes based on what I'm describing. The only difference is that since an artificial intelligence is capable of going through these combinations in quite a quick order, it is able to effectively optimize your tax reform based on existent numbers in your accounting software whether it be QuickBooks or it be Excel. Knowing which numbers are associated with which categories, artificial intelligence can iterate through the different combinations and find the one that benefits you the most.

Years of Information in Minutes

The best part about building an artificial intelligence to do this sort of system is that that artificial intelligence will have access to tax information for years. Instead of needing to keep your tax information on file and ready for them in paper format, your system will be able to consistently retain the tax forms made in previous years.

In addition to this, you can actually utilize this information to a further point because unlike market research or product research, tax research allows you to see where you could expand your benefits to take advantage of government supplements. The benefit of having years of tax information is that when new tax laws are passed, some of them are retrograde because they want to support foreign companies coming to their soil. Therefore, a country with a lot of retrograde supplements will benefit foreign companies that come there without expectations of supplements in the future, but those companies will understand there will likely be retrograde supplements in the future while they stay there.

Associated Searches

Another thing that having an artificial intelligence that simplifies tax code can do is allow for associated searches. Usually, if you are looking for a particular portion of the tax code you are very likely to also be searching for things that either that tax code depends on or relative dependencies that rely on that specific tax code.

As you create a list of dependencies, you can then create a list of associated searches so that if you are looking for one item you can find the relative items that are associated with that and save on time. In addition to this, when using tax data to run analysis on that tax data you can then pull up associated data to create links that would not have normally been detected by artificial intelligence because the relationship is only defined in tax code and not mathematically. The situations in which this case is true is quite numerous so being able to set up associated searches saves on time for filling out tax forms as well as performing research on that data.

Law Differentiation

A unique thing that happens in some of the countries that you might have try to settle is the differences in laws that are followed by the provinces of those nations. Japan is a country where there are multiple provinces and something as simple as the age of consent varies widely between the provinces. You have some of the provinces having this age of consent be nearer to the age of 20 while in other provinces this age is nearer to that of 10.

The reason why this is important is because understanding the differentiation in laws per the provinces of the government helps to establish what can work in that province versus other provinces. There are entire organizations built around the concept of using the laws to find the best places for companies to expand and make their products in. This is because of the wild variations of the laws that go into practice per province versus the laws that are passed by the government. Artificial intelligence is able to pick up on words that differ from each other in the tax code relatively easily.

The way that a human would be able to detect the difference in the tax code is usually by simply reading the two definitions side-by-side and then making distinctions of this difference. This is a simple step that is very repeatable, which means that artificial intelligence can handle this and create databases specifically for handling provinces and the differentiation of law in the provinces. This saves a massive amount of time when it comes to noting the tax simplification for your company.

Reduces Administrative Time and Cost

Monitor Levels of Safety

Part of the reason for compliance is to ensure that safety in the workplace is taking place. Having an artificial intelligence monitoring every aspect of the facility ensures that this safety follows the level of compliance that it needs to follow. Most facilities require a daily inspection to make sure that things are up to par when it comes to safety standards. Not all of the workers will put on the safety gear that they agreed to wear on a daily basis, not all of them will operate the machinery perfectly, and it's just a normal human error that causes a failure to meet safety compliance. By having artificial intelligence monitor everything that goes on in this facility, the company can ensure that it always compliance and that he can prove that it was not the company's fault when it came to injury on the workers behalf.

Lower Accountant Count

By being able to simplify your tax forms and being able to automatically convert current financial information into tax submittable forms, you ultimately lower the number of accountants and financial departments you need to rely on. You cannot completely remove accountants from the equation simply because artificial intelligence is not perfect and since it is built by humans, it is prone to error. It is far better to have an accountant ensure that the tax form follows the necessary tax codes, but the majority of the work can be done by artificial intelligence, you just need to modify it when it doesn't do the job correctly.

Predicting Consumer Behavior – Conjoint Analysis

The Obvious and the Not Obvious

Keep Track of Overall Product Sales

Consumers are predictable, which means that they are usually very good at differentiating between products they want and products they don't want. This means that if a product is failing, it's going to be rather obvious when it is compared to other products.

A marketing manager or a product manager is paid to figure out what products are selling, and which ones are not. However, these steps to figure this out are long, tedious, and repetitive. Knowing that it is repetitive means that artificial intelligence can handle it.

You simply use the same equations the managers would have used and plug in the necessary input. The only difference is that you can now performs additional calculations.

Keep Track of Sales Associates

The old marker of a good sales associate used to be how many products they could sell, but new information suggests that if you want consistent buyers then you have to maintain happy customers. A good sales associate can sell anything, but that doesn't mean the customer is going to be happy with their purchase. Using a survey along with artificial intelligence, a company can keep track of how a sales associate has an impact on user happiness.

This aggregate data can then be utilized to understand the level of skill each sales associate, identify which ones are lacking behind, and find out where they lack in terms of their colleagues. It is important to know this because you can improve coworkers by matching up those who are lagging behind others to the ones that achieve the best scores.

Differentiate Attributes

Both products and sales associate performance improvements boil down to the same terminology; differentiating attributes. There are some attributes that harm and some that benefit. As a company, it is your job to figure out how to minimize the harmful and maximize the beneficial.

Conjoint Analysis

Attribute Analysis

The purpose of a Conjoint Analysis is that it allows you to measure what consumers choose as your most successful products or associates based on their attributes. Conjoint Analysis can be done by hand but there's a reason why machines and artificial intelligence are used most of the time. Machines are used because the calculations can be done quickly, consistently, and accurately every time. If you were to have a storefront as big as Amazon, it would be impossible for a single human to run this analysis by hand on all of the products that exist.

Artificial intelligence can do the analysis in real-time and predict better versions of the more preferred products. If you have a black phone with a 24MP camera and a green phone with a 12MP that are equally profitable, it becomes difficult to know why they are popular. Artificial intelligence is able to utilize known sales information and create probabilities that point to the reasons for why. For instance, it might find that more green phones are sold in-store while the black-phones are sold online. You would have to go through every receipt to discern this reason for yourself, but artificial intelligence can automatically do this given the right parameters and inputs.

Product Variations

It used to be that you could purchase smartphones in a variety of different colors and design. Over the years, companies of slowly reduced the amount of colors and designs to their most preferred versions. The problem with selling products is managing the variations in that product line and these algorithms that utilize conjoint analysis, K cluster and other algorithms allow you to refine the variations to the most profitable versions of themselves. Artificial intelligence is able to take massive data and create attributes without needing labels, but it can also handle and categorize data and it really just depends on how you build it. Artificial intelligence can lead the way to more profitable products by analyzing which product variations are the most profitable and predict new, even more profitable variations based on that data.

Finding Consumer Preferences

By building artificial intelligence that's able to predict more profitable variations of your products, you are able to better understand your consumer preferences. This becomes a cycle where are you make a product that is produced based off of data your artificial intelligence provides you, consumers buy that product, and the variation that they buy the most often of is the next base for your new product in the future. Therefore, if most consumers buy a smartphone that has a fantastic camera but low storage and a lower amount of them buy the smartphone with the high storage but the low-quality camera, you begin to understand that your consumer base is more concerned about camera quality and so you can improve the camera without improving the storage to get a similar result.

Digital Customer Acquisition Tools – Online ads and statistical tools

Conversational Marketing Like Drift

Segment Marketing from Sales

This type of software is really meant to remove the confusion from marketing emails and sales emails. To give you an example, if you have a marketing campaign going on that invites users to the platform to take advantage of a specific coupon and sending out these emails, if the sales department doesn't effectively communicate that coupons are being used then that customer might try to reuse that coupon. They then go through the website and through the process, only to find out they cannot use the coupon because they've already used it. This creates irritation on the customer side and will likely demand the coupon be taken via customer service otherwise they shouldn't have sent the email out.

Additionally, this type of tool is really good for small businesses that need to consolidate their management space. This tool provides the company with the ability to manage marketing and sales in the same location without getting the two confused.

Premade Strategies for Retention

Since the platform needs to invite you back as a customer on a regular basis, they have to provide more than just removing the confusion between sales and marketing. Therefore, they have performed significant research into strategies that will help retain the user that you do not need to create on your own. There's a significant amount of research that goes into making a retention strategy and they go one step further by providing you with default strategies you can use that have worked in the past.

Account Based Responses

The best part about this is that this platform is really for marketing based on different account levels or attributes. If an account is inactive for a certain amount of time, you can set up a marketing campaign to bring them back to using your tools. If an account activity

In a specific area, you can then market the higher subscriptions that will

benefit those areas in specificity. This is what's known as account-based

responses, where the marketing is based on what is currently happening

with that account. This is far more precision in market targeting than

any other practice and therefore usually sees the best results.

Landing Page Conversions like Unbounce and Clickfunnels

Landing Page Creators without Code

Normally, a landing page requires a few things for creation. It

normally requires a front-end designer that will design a landing page

and it also requires a marketing specialist to determine what needs to go

on that landing page. Usually, you can acquire this but the front-end

developer is another section you need to create a landing page that will

go on a website. You then need a developer to implement that landing

page into your website so that it works and usually need to have them on for at least 3 months to make sure that the landing pages working the way you want.

The benefit of using services like these means that you are able to create landing pages without investing that much money into the creation of the landing page. You can simply choose the one you want to use and fill it with the content you want to use for that landing page. This is far simpler than the old process of creating landing pages, and it's usually even better because they are landing pages that are designed to have known conversion rates of a high amounts.

Working and Applicable Themed Creation

Best yet, services like these specifically categorize landing pages by themes such as what industry you plan to be selling to. This means that you can create specialized landing pages for different industries that you plan to sell to so that you get even higher conversion rates. This allows the themes to be applicable to practically any industry that you plan to sell in.

Onsite Retargeting like OptiMonk

Engage Customers Leaving a Website

If you rewind by around five years, you are likely to find that the idea of engaging customers when they are leaving websites is somewhat of a new idea. Now, it's almost commonplace to see that websites want to prevent you from leaving so that they are able to go through the sale.

Services like these are primarily geared for ensuring that you still are able to sell to a customer that is either leaving a website or has already left the website. Usually, the reason why a customer visits your website is because they are interested in the product you plan to sell them. The problem is you don't know why they don't want to stay on your website and make a purchase, which this allows you to further inquire using email marketing or reminding them that they might get a 10% discount now that they have tried to leave the website. In either case, you are able to follow up with the sale and continue to sell it to them.

Create an Email List for Difficult Customers

As I just mentioned, you can create an email list based off of customers that try to leave the website. Normally, it is best to combine this with a free teaser gift for allowing you to collect their email address. What this does is it allows you to have access to customers that you would have previously lost and had no way of knowing why they didn't purchase anything.

Recover From Cart Abandonment

Sometimes, a customer simply needs to do something else on a different website that is not on your website. They had already planned to buy something on your website, but their distraction led them to leave their cart abandoned on your website. You can use an email or a pop-up reminder to remind them that they have items in their cart that they haven't bought yet. This is important because you may end up being able to sell to that customer simply because you had a service that reminded them.

Referral Software Like ReferralCandy

Encourage Customers to be Advertisements

Services like these allow you to promote your service through the customers that come to your website. Usually, the way that you lure customers to your website is with a free gift of some kind. When they land on the website, they can then submit their email and maybe make a purchase. Once they submit their email, they will get a referral and they will be notified that they can get a reward if they lead their friends to your website. This strategy is so effective that it has been around for nearly a decade by now and it will feed customers into your website. The best part about this is that like the other items, you simply plug it into your website and it begins working and is a separate component from the website.

Accruing an Email List

By utilizing this tool, you then begin to accrue an email list, but it is an email list of people who are willing to be interested in your products. You see, the problem with referral programs in the past was that you could then go on coupon websites and use coupons that were found there to make purchases when you would not normally have access to the coupons. Essentially, this would effectively exploit the weakness of the referral program and a bunch of fake accounts would be created to try and take advantage of your referral program and reverse it to benefit the individual attacking. Software like this protects you from this by using artificial intelligence to filter out real users from fake users.

Theme Based Build and Referral Promotion

Like the previous services, this type of service usually comes with pre-built themes that have already been vetted for the market. That means you don't need to worry about the design of the website because it's already something that works, but it will actually guide you through implementing such a system so that it looks good on your website rather than you just fumbling about with different color themes.

Analytical Tools Like HotJar

Record User Activity

Sometimes you don't just need aggregate data to understand what happening in the website because sometimes you want to see exactly what users are doing. A service like this allows you to actually record the actions of individual people's mice as they traverse your website. This gives you a very deep understanding of the process and patterns in which your customers are interpreting your website.

User Activity Analysis like HeatMaps and Scroll Percentages

This website still uses the artificial intelligence to provide you with the necessary tools to make a proper analysis on user activity for your website. It has numerous tools such as heat maps and scroll percentages. A heat map is an image that runs an algorithmic process to find the most visited points on a web page per web page. Scroll percentages help you understand just how far the user gets as they scroll down your website to determine what content they are skipping out on.

Feedback Mechanisms

Perhaps the best components of a service like this is that you can run surveys and polls on the fly for your website. You can get instant access to feedback that would have otherwise needed to be built into the website for it to happen at all. With this service, you simply go in and create the survey or poll that you want to come up on your website and then you can fire it based off of an event. This allows you to receive feedback for several different components of your website from various different sample groups.

Conclusion

AI Shouldn't Be Feared

Artificial intelligence is often underneath the eye of scrutiny because of how much information it requires to be useful. For a long time, many people have harkened the visage of artificial intelligence to that of the Terminator movies where they eventually get so smart that they annihilate the human race. Artificial intelligence is nowhere near that because as we can see with Google Voice, the number of problems we have to correct with a simple language recognition program should prove just how far from this point we are as a society. We have several different implementations of artificial intelligence but almost every single one of them is rudimentary compared to the necessity that is needed to make it a threat.

AI Brings Out the Best of the Best

Artificial intelligence is a low-cost investment that ultimately brings out some of the best results that money can buy. You only need access to the mathematics and the data for artificial intelligence to become useful. This means that any business from the lowest level to the highest level you can get the same high-quality data and predictions needed to propel their business forward. Artificial intelligence, rudimentary as it may be, ultimately benefits us as companies and as a society as we push forward with advancements into this new technology. Don't be afraid to use it and get the advantage while it's still an advantage.

BLOCKCHAIN TECHNOLOGY REVOLUTION IN BUSINESS EXPLAINED

Disclaimer

The history of blockchain and cryptocurrency

Before we get started, I just wanted to talk about how excited I am about the potential future of blockchain technology; not just in business, but in every aspect of our daily lives. For the last decade, we have seen technology progress at breakneck speed. The advent of the internet, smartphone technology, video streaming, social media and online businesses have changed the face of the planet. However, there have been some downsides. Online privacy has taken a backseat for the average consumer, in the search of more business. Google and Facebook have decided that the pursuit of more clicks and revenue supersede our personal privacy.

Enter blockchain. Blockchain has the potential to give us more privacy through innovative encryption of our online activities. This could give businesses the info they need while protecting our privacy.

It also has the potential to revolutionize the way that transactions are made; eliminating the need for the middleman to check the authenticity of these transactions.

Now let's get started!

The Original Problem with Digital Transactions

Most Digital Purchases are Bank Transfers

Believe it or not, most digital transactions still occur like a normal bank transfer would nearly three decades ago. The only real difference is that it goes to a middleman before it is approved. You start a transaction since you want to pay for some item, the amount or, rather, the number associated with the amount is then sent to an auction clearing House. This clearing House talks with the associated bank you're giving the amount of money to and ensures there is an account to transfer to. That same clearing house also asks if the bank you're transferring from has an account with the number of the amount associated with it. Once it confirms both sides exist and you have money, it then commits the transaction.

As computers have gotten faster, these transactions have gotten faster as a result and so it's barely noticeable to the average consumer. However, essentially, you are doing a wire transfer every time you use a card, every time you use PayPal, and every time you use something like Google Pay.

Bank Transfers Are Tracked with Physical Items

The way that banks are able to do this is because they have an exchange rate. That is, they have a rate of which items can be exchanged in their program otherwise known as Fiat money. It is not the same exchange rate that is referred to when talking about the value difference between dollars.

What this is talking about is that the bank in question has a certain amount of value in dollars that can be transferred.

While money used to be a physical item, it's not anymore, well with a majority of money that there is. Most money is digital because it allows for inflation and deflation of a currency by those who make that money. This actually makes the value of that money much easier to control rather than relying on physical items, but banks still make transfers with physical items.

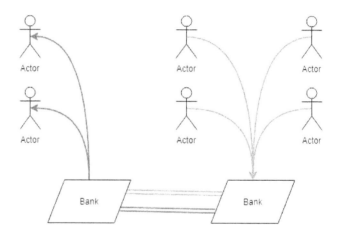

Usually, at the end of the week, there is a security truck that transfers the amount of money that has been transferred out of that bank going to a depository that will then disperse this money amongst the banks who have claims to it. The bank that sent out the money will also get its own version of a security truck holding money for money that was transferred to that bank. Physical, paper money is still in circulation because of the system as it represents a permanency with the money. You can't copy and paste a dollar bill… well, easily that is.

The Double Spending Riddle

Digital transfers have actually been around for much longer as a concept than they have been in practice. This is because digital currency is a very easy and international way of purchasing items. There was just one problem with doing everything on the computer and that was because of an incredible invention that most of us use in our daily, if not weekly, lives; copy and paste.

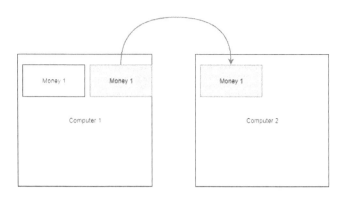

When you made a transfer of digital money, how would you go about preventing the ability to copy and paste that money just so you could have a hundred more like it? This was known as the double spending problem. In the beginning of computing history there was virtually no way of preventing the double spending problem and it is only through the internet and encryption that we can truly make our way past this problem, for a temporary amount of time.

Banks Are Prone to Double Spending, You Just Don't Know

As I noted earlier, you can do a form of copy and paste with Fiat money. Money is just an imaginary thing in that the dollar bill in your hand represents a value everyone created and doesn't actually have a measurable material value beyond the cost it took to make it. Essentially, everybody in society said that this dollar is worth this amount.

Therefore, if you can make something that looks identical to this then you have a double spending problem; it's just not digital. We often refer to this as fraud as it is a term used for trying to lie about something, which means it could be money fraud, wire fraud, or bank fraud amongst the many other types of frauds. Knowing that money could be copied and pasted, those that printed the money took steps to ensure that doing so was incredibly difficult. This is why you have things like special color ink, ridges around the edges, and custom fonts on pieces of cloth as this makes it difficult to copy it unless you have the original printer. Essentially, the first monetary form of encryption.

How to Prevent Double Spending?

Everyone Keeps Track of Everyone

Now, the first thing about the double spending problem was that they needed a way to ensure that double spending couldn't happen between two parties. Let's say that you decide to make a contract with a company for certain amount of money and then you make a contract, again, with that same company. They are the exact same contract; the only difference is that you get paid twice as much for the same amount of time. How does the company protect itself from contract fraud in this case?

There is actually a position for this and it's known as a notary. A notary has the job of being physically present at the time of signing a contract so that they can say they witnessed this contract happen and then the second contract is invalidated if that notary is not there to sign for the company. Therefore, if you manage to trick a second employee to sign a contract with you, the notary who was there for the first contract signing will know that this second contract is attempted contract fraud. This is essentially how you prevent double spending and we'll get into the details later on.

Worker ID

Now, most systems that keep track of contract invoices will usually have a form of identification. Work at a place long enough and they will give you your own specific workers identification number and this is simply to identify you as the person who you say you are. You often see this in security card keys that are way too vulnerable for most businesses.

```
sa064t1aMPjmBs2Cd2v-
m69YbFmMvpfNt9kt6qC3PF0l8eSV5nof5Yn2hCmS
```

A cryptocurrency coin has a form of this and this is often known as the worker ID (seen above). The worker ID is often a combination of your specific ID attached to a randomized number that represents your worker on the computer that got the money for you. When you make the transaction, the portion of the worker ID that represents you is actually switched out with the worker ID part that represents the person you're transferring too. This is how identification of coin holders happen.

Coin ID

```
sa064t1aMPjmBs2Cd2v-
m69YbFmMvpfNt9kt6qC3PF0l8eSV5nof5Yn2hCmS
```

Identifications of coins is a little bit different because the identification number or stream is actually given to you by the algorithm that runs the entire system. This is a mostly randomized stream of letters and numbers partly because there are already coins in the system and those coins clash if there was an identical coin with the same stream. The only difference is that the algorithm that creates this also utilizes techniques so that the coins cannot be created in a linear order but rather an order that matches the function of the algorithm.

Therefore, when you get a coin it is a combination of randomized letters and numbers as well as a sequential combination of those in alignment with the algorithmic function designed to further randomize the coin but in a randomization pattern that can be reversed. This is how coin identification numbers are created so that no two coins have the same identification number attached to them and no one can figure out the next coin and just create the coin on their desktop.

The System Isn't Complete

The system itself is not very complete and this is why you have so many versions of cryptocurrencies out on the market because there's always someone that either wants the same amount of success they saw out of Bitcoin or someone thinks they've got a better idea for a cryptocurrency. Every year there is an improvement on the technology that exists in this market simply because it is profitable for it to do so.

Not only that, but the system is also still very vulnerable, and this isn't necessarily because of the mechanisms in place used to identify coins and workers. The vulnerability of the system is a two-part problem dealing with encryption and a difference of ideas. Encryption is used at nearly every level until you get to the point where you can earn the coin in the methodology of how that coin is earned.

Encryption is Needed for This to Work

Anyone Can Transfer the Number 1

The primary problem with the double spending problem was the fact that it could be copied and pasted. The easiest way to solve this problem is to simply make it so that it is almost impossible to copy and paste it. You can still copy and paste a blockchain node but, due to encryption, even if you were trying to use it would invalidate the use beyond a single transaction.

Cryptocurrency is not tied to the device but, rather, tied to the encryption of the coin. The encryption algorithm is unknown to anybody that is not the creator of the blockchain. This means that all that needs to happen is the blockchain needs to verify the previously recorded encryption set for both the worker ID and the coin ID. If these are the same, then what happens is that when you transfer the coin the worker ID is changed to the new identified worker. This ID is attached to a wallet and the wallet contains the worker ID that is stamped on every coin that goes into that wallet.

This means that if a person is using a cryptocurrency coin in their wallet and that coin was forcibly placed in their wallet, the blockchain would compare the worker ID to the wallet ID and deny the transaction because they didn't match. This means that if you try to have a fake coin attached to your wallet ID, the same transaction validation will occur. The coin ID will be compared to what is in the blockchain and the worker ID will be compared to the worker ID that's in the blockchain and when the blockchain sees that the worker ID attached to your coin does not match the worker ID in the blockchain, it will deny that transaction.

Breaking and Gaming the System

The problem with this is that encryption is not infallible. As an IT security specialist often says, security is not an impregnable defense but rather a delay tactic. Most blockchains will use an encryption method that will take years to break unless they know the specific passphrase needed to translate the letters.

This brings up the issue of vulnerability because what happens if the person who made it is trapped and forced to give their own passphrase? Normally, the best security against this is a rotating password that is based on a randomization method so that not even the creator will know what the passphrase is without giving himself or herself direct access.

How the Ledger Works

Everyone Sees A Transaction

The first thing to know about why this blockchain is so secure is that everyone that participates in the blockchain has a copy of the ledger. Everyone in the network can see a transaction but because it would be costly to make potentially billions of transactions every second over a network, a random

group is chosen to compare the transaction to other ledgers.

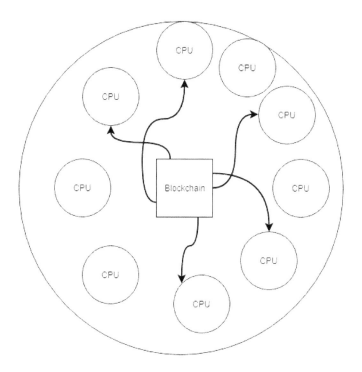

The network chooses currently participating devices to

borrow their portion of the ledger so that a transaction can be

compared against it.

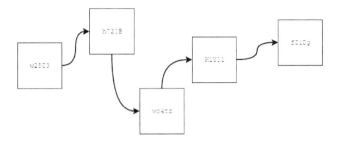

Every time a transaction is recorded, it becomes a block in the chain and thus you are able to quickly search for a transaction that happened with a specific coin.

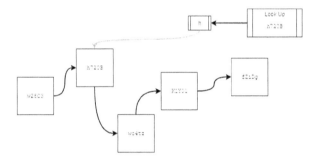

This makes the search algorithm extremely fast and then also makes it so that the comparisons can also happen at a very fast rate. Having the choices randomized increases the likelihood that no one will be able to tamper with the ledgers so that a transaction can pass, which further decreases the likelihood of fraud.

Transfers Are Verified with Ledger Comparison

Everyone on the network has their own version of the ledger making all of the ledgers original. When a transaction occurs, an update is sent to all the devices connected to the network so that everybody has the same ledger even though everybody has an original ledger.

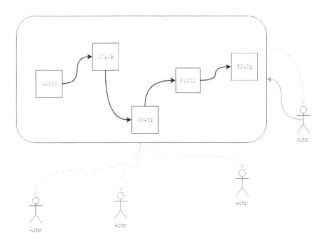

This is so that everyone can become a target for verification. 5 to 20 people on the network are chosen to see if the coin belongs to the person trying to trade it and to see if that coin has ever been traded to that person by the network of the cryptocurrency or by anyone else in the system that eventually got it from the network. Based on the percentage, the network decides whether a transaction will go through or not.

It Usually Has A 100% Threshold

Most networks will have a 100% threshold before finally approving a certain transaction. There are some optimistic networks that allow transactions that happen at a lower threshold, but most networks try to use 100% thresholds to prevent this fraud.

The worker ID is checked to see if that worker ID has ever been on the network and is also known as *your address*, the coin is checked to see if it was ever issued by the network, and then the coin ID is compared with any worker ID that had made a transaction in the past to see if it connects to the worker ID that's making a transaction.

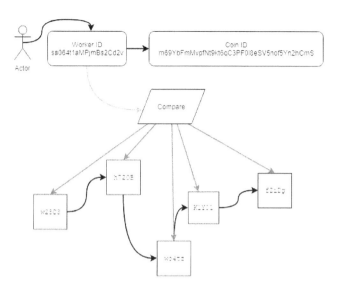

This is actually essential for understanding what blocks
are in the blockchain. The most basic of blocks are simply a date
of transaction, a date of verification, worker ID and a coin ID.
This is also important to understanding the efficiency of the
blockchain.

Blocks are Used for Small Comparisons

That was needed to understand the efficiency of the
blockchain because if it wasn't built like this, it would take
forever to make transactions. Millions of people interact with
the blockchain which means that if you built a blockchain to

simply look through all the available records it would have to go through millions of records, really a likelihood of sextillion of records at this point, before it ever reached the record you were trying to compare two.

Instead, you have a hierarchy system. Perhaps the first element would be the worker ID itself because it is the smallest number that you have.

sa064t1aMPjmBs2Cd2v

Then you have the coin ID, which is not as small as the worker ID but it's still pretty small.

m69YbFmMvpfNt9kt6qC3PF0l8eSV5nof5Yn2hCmS

By using a worker ID, they can issue numbers in a linear pattern and this allows for the binary search.

```
1    require 'benchmark'
2    number = 10004
3
4    def binary_search(number)
5        blockchain = (1..1000000).to_a
6        binary_term_right = blockchain.length
7        binary_term_middle = blockchain.length / 2
8        binary_term_left = 0
9        guess = true
10       passes = 0
11       loop do
12
13           binary_term_middle = binary_term_left + (binary_term_right - binary_term_left)/2
14
15           if blockchain[binary_term_middle] == number
16               break
17           elsif blockchain[binary_term_middle] < number
18               binary_term_left = binary_term_middle + 1
19           else
20               binary_term_right = binary_term_middle - 1
21           end
22
23           if passes > 70
24               break
25           end
26           passes += 1
27       end
28       puts "This took #{passes} passes."
29   end
30
31   Benchmark.bmbm do |x|
32       x.report { binary_search(number) }
33   end
```

A binary search will almost always take several low loops.

```
Rehearsal -----------------------------------
This took 19 passes.
   0.047000    0.000000    0.047000 (   0.056323)
------------------------------------ total: 0.047000sec
```

While the algorithm above uses an integer, we cannot

know how the developer chooses to convert letters into numbers

as this may be a part of the conversion, so they may even use

ASCII numbers, which would be easy but not secure. This

means that the hierarchical search pattern only needs to go

through a ledger seven times to find a worker ID. Then, since

193

there's a very small number of coins associated with most worker IDs you can just run a standard loop. The difference between this way of searching and the previous way of searching is that you are skipping trillions of instructions every second. This is why the blockchain is in nodes rather than a linear structure because it allows coins to be attached to worker IDs in the lookup system and the strings only need to be compared for their real value with very basic calculations on top of that to determine the validity of the date. This is how the blockchain works.

How It Spreads

Rarity Equals Value

The problem with this technology is that you have to spread it first and while this book will flip back and forth between cryptocurrency and blockchain, there needs to be an understanding that for a very long time blockchain was solely used for cryptocurrency and this was/is how it was/is spread.

In the beginning, Bitcoin was worthless and it was the equivalent of the fake money you see in games used for microtransactions. Essentially, Bitcoin was given away for free but eventually, people had to mine for Bitcoin and they could no longer get it for free. They had to participate in the transactions of Bitcoin to further their amount of Bitcoin, which drove up the minimum requirement of work to get Bitcoin and thus the value began.

Competitiveness Drives Rarity

By creating a system that allows people to earn money by facilitating transactions between users, you create an economy that profits off itself. By setting a finite amount of coins that can be earned you also increase the rarity of those coins.

By creating a system for people to earn money you invite competitiveness to earn more money. This has resulted in the insane value of Bitcoin and why it dropped so drastically once other cryptocurrencies started to gain a foothold in the market. Bitcoin, itself, is actually really unstable because it's imaginary money and it only has value to those who give it value.

Therefore, when the complex equations get so difficult that almost no one can mine them, and no one can really make money off of them anymore, the value drops. It was predicted in the very beginning that this would happen, but a lot of people tend to not listen to warnings as shown by the many customer support help desks that come with almost every company. Knowing this, people began to look at the intrinsic value of Bitcoin because even though it may lose its value, its technology was like nothing we had ever seen before. However, cryptocurrency as a category will always be worth something now that it exists and here's why.

This is How Fiat Currency Works

Cryptocurrency is similar to Fiat money except that Fiat money usually has a physical item representing it such as the US dollar bill. The value on that US dollar bill is imaginary and it is only ever worth the number we prescribed to it. Therefore, in reality, we could look at a US dollar bill and say that it was worth $100 in our store while a $100 bill was worth $1. Now, there are some rules to protect against this. However, if those rules didn't exist and we chose not to follow any new rules then we could very well say this was the reality of the dollar bill versus the hundred-dollar bill even though it makes no sense because 100 is supposed to be more than 1.

However, that is how Fiat money works. There used to be something known as the golden standard and this was a time when the dollar was not just a numerical amount but exact representation of gold contained by the US Treasury. This gold standard used to be how many countries traded items in the past and now money is just an imaginary number attached to a piece of clothing that people call paper that represents the amount of work that you did to get that money. Cryptocurrency is the same thing except it's purely digital and that's it. Cryptocurrency has security implications and money has its own security implications such a special ink and unique font types. Just like cryptocurrency, money also has a chain of verification attached to it and it's usually performed by people who have the tools to verify that you are trading a U.S. dollar. This was caused by the dollars own double spending problem because there's nothing to say that you couldn't just have a printer copy and paste the dollar bill. Therefore, cryptocurrency will always have a value because it is a form of Fiat money and there is an economic society that supports it.

Cryptocurrency Only Works with the Internet

However, the one huge flaw with cryptocurrency is the requirement that you need the Internet for it to even work. You need the internet to obtain the cryptocurrency, you need it to trade it, and you need it to provide validation services. This is the one huge difference between money and cryptocurrency because money can be printed offline and you don't actually need the Internet. It means that if ever there comes a time where cryptocurrency represents the GDP of a country, the internet of that country will become the critical infrastructure that supports that country's economy.

Cryptomining Is A Collection of Methods

Proof of Work Methodology

The most common way that the blockchain was spread on the internet was with a concept known as proof of work. People would allow their GPUs to work on some algorithmic problem generated by the blockchain itself and when those problems were solved then it represented a valid transaction. By solving these problems, people would earn a percentage of the transaction that occurred as a reward for doing the work. Solving the algorithm in producing the correct answer would serve as the proof of work having been done and thus the proof of work concept was both born and named.

Bitcoin primarily is responsible for making this optional method of obtaining cryptocurrency the most popular on the market. At first, it wasn't that difficult to do this type of algebraic quick solving and so a lot of people experienced a very low bar entry into the game. This is actually what happens when someone says they're mining for currency, they are using their computing power to solve algorithmic problems.

CPU Mining

Now that you know this, it was quite obvious that CPUs would be the first that took the brunt of the calculations. This reasonably set the bar to the highest amount of computing power that a CPU could output and so, really, it generally set crypto miners apart from regular people who bought computer components.

The obvious choice was to use server cores as the number of cores gave you more options to multi-thread and solve more problems. You could generally make quite a bit of money off of this. The only problem is that even the power of CPUs wasn't quite capable of getting and succeeding above others who had already been earning money. In other words, no matter how powerful the CPU was there was a bottom line for how much you could make and everyone joining that market made everyone else in the market lose money. The calculations themselves were rather simple, it was just a matter of complexity that the blockchain had chosen beforehand that was the problem. There wasn't an efficient way of breaking down the problem into smaller chunks until there was.

GPU Mining (Graphics Processing Unit)

Lo and behold the world of crypto mining entered the GPU market thanks to a handful of programmers. You see, these programmers had found a way to allow GPUs to make the primary calculations for the algorithm and the GPU market had a lot of potential in it. Essentially, the graphical processor unit was responsible for producing the visual aspects of a computer. Almost all motherboards come with a graphical processor unit by default, but graphics cards are far more powerful and even integrated graphics are more powerful than the small graphical chips that provide motherboard BIOS interface. Instead of having 32 or 64 cores working on an algorithmic problem, you could have somewhere around 1000 to 2000 cuda cores working on that same problem. This was an exponential leap in crypto mining and it allowed a lot of people to gain a lot of Bitcoin really fast before everyone figured out what they were doing.

Now, the Bitcoin game is nearly at its end and very few people are left mining the cryptocurrency that is Bitcoin. The reason why is because of a couple factors. You have GPUs only being able to earn so much and only being able to run at a certain wattage, so you have a set amount of money you can really make. Recently, bigger companies have started entering the crypto game. Competing with giant farms with thousands of cryptocurrency mining cards is not really viable for the average person. Mix this with trying to buy computer components on the market. For a good year, gamers and PC enthusiasts as well as companies who regularly update their computers had to fork out tons of money just to upgrade their GPU. As of writing this book, now that Bitcoin has drastically dropped in value, the mining cards are flooding the market and overtaking the general sales of brand-new graphics cards. This is because these crypto mining farms are often selling to make a little bit of profit back for their investment. Many people have succumbed to buying a mining graphics card only to find out that the graphics card underperforms because the mining farm changed the default settings on the device to make it more compatible with mining. Essentially, the proof-of-work concept had finally worked itself to death.

With Each Coin Comes A Different Type of Blockchain Sort Of

The BlockChain could Only Be Used for BitCoin

In the beginning, the blockchain didn't really have a purpose outside of Bitcoin primarily because the original creator only thought of the blockchain as it related to Bitcoin. This is a common thing that happens with anyone working on a project and it happened to the creator of Bitcoin. The creator of Bitcoin was so focused on developing a cryptocurrency that worked and fixed many of the problems of the past. They weren't exactly focused on creating a revolutionary technology that could change nearly all industries.

It took a very long time for even programmers to get used to the concept of the blockchain and even longer for cryptocurrency to evolve. As explained, cryptocurrency mining usually happened on the CPU at the very beginning. This is because no one really knew of a way that the complex calculations of the algorithm could be performed on the GPU. After a few updates of Bitcoin, more people began to understand it and higher people in the chain of technology began to understand what the crypto-currency market was capable of. They had the experience with developing GPUs. They also had the analytical skills to mine cryptocurrency by tricking the GPU into working together like a CPU.

Understanding the Blockchain Caused Comparisons

Once the technology got out into the market, people slowly digested what the blockchain was and this allowed people to begin to compare the blockchain to other items that might be useful. In other words, the blockchain wasn't useful at first until people started thinking about situations in which having a way to verify information was ideal in a decentralized manner.

While a blockchain could serve as a form of identification, that information also had to be able to either change or be replaced and so the idea of something like smart contracts came to exist. With a smart contract, you're able to set the properties of the contract and then if you need to change them or modify them you can access that contract, make changes, and put it back. The only difference is that when you put it back, it's now at the end of the blockchain. So, the changes to the contract are recorded in the blockchain. While the technology is not 100% perfect, it is significantly better than the current technology that we have.

The Smart Contract Was Born

Once people could realize that you could use a blockchain to store information, they wanted to make a way that allowed you to store information inside of a cryptocurrency coin. They would take an old idea and the security of that old idea and repurpose it so that it is very quick to market with possible changes in the future.

There were plenty of alternative cryptocurrencies on the market because whenever you have a successful product, you will tend to have competitors that want to capitalize on the money that's being made in that situation. The reason why smart contracts are different is because it was on top of a new cryptocurrency. It still used a very similar blockchain to previous cryptocurrency while also adding functionality to it.

Most of this technology is tied to the cryptocurrency that was invented for it because in the cryptocurrency world, the most common way you can spread the new technology is by having people invest in that technology. Thus, once Etherium came out on to the market, that was when the smart contract was really born.

The Smart Contract Wasn't Directly Connected

Building your cryptocurrency off of previous models and adding functionality to it presents a unique challenge of itself because the blockchain wasn't it really supposed to be interacted with. The fact that the blockchain wasn't supposed to be interacted with was the primary security feature of the blockchain itself so how do you go about making a non-interactable object blockchain-compatible?

The concept of the smart contract is rather confusing to most people because they're very used to understanding the blockchain. You can't actually touch the blockchain because the machine itself touches it. Knowing this, it is difficult to understand how you go about creating a smart contract, but a smart contract is really an extended coin.

Let's talk about internet packets because it's actually very easy to understand a smart contract once you understand internet packets. When an internet packet is created, you first have the source of the internet packet, the destination of that packet, the lifetime that packet will have on the network, and then a small chunk of information representing your content on the network and usually these packets are broken down into very small chunks with a sequence marker. A cryptocurrency coin is made up of a worker ID that represents your wallet, the coin ID and then the last part, which is the smart contract, is anything attached to that coin ID.

Therefore, when you make a smart contract you are really making a cryptocurrency coin but you are adding everything in the smart contract at the end of a coin and so you have a unique coin ID for every smart contract that there is. This is what they mean when you are not directly interacting with the blockchain itself in a smart contract because you are fulfilling the requirements to put content after what would normally be the coin ID. Sometimes the contents are the coin ID itself but usually, it's not because you want it so that the system is able to mix your content up with a coin ID so it's not easily breakable within the system. Human language has been the linchpin that destroys many encryption methods because of common letters and phrases that we use so you want to add on to the coin a little bit of string that's randomized or inside of the string to cause disruption in the encryption so that not everything is perfectly aligned. However, while that's getting off topic, the way that a smart contract works is that it adds the information at the end of the coin and represents the coin itself and so when you submit it to the blockchain you are really still submitting a cryptocurrency coin.

The Ledger is The Blockchain

Ledger is Used By Rationalizers

The terms used to describe cryptocurrency and its methodologies are confusing. They are not confusing for the reason that you think. They begin to make sense once you have a background in programming and a small amount of knowledge in Fiat money accounting. Block is actually a programming term that refers to any line of code that means more than three lines of code in order to execute. This could be two lines but usually, it's three as a minimum. This is used as a way of describing blocks of code inside of programming and then you have the inline code and that's just code that takes up a single line.

So, if you really think about it, it makes quite a bit of sense because all you really need to see it as understanding what a chain mechanism is as well as an understanding of what block means in programming. To put it bluntly, a blockchain is a chain of code blocks. This represents the ledger that many cryptocurrencies use. The reason why it's so secure is that once a block has been made in the chain, the block cannot be modified. To modify the block, you have to modify to modify every single transaction in the chain of the block.

So, basically if you hear the term ledger, we are actually talking about the blockchain.

We just went through the basics of blockchain, including its history; and how it has evolved to what it is today.

How BlockChain Technology is Implemented in Finance and technology

In this chapter, we're going to go through a number of different applications of blockchain in finance, technology and programming.

Blockchain Anon Systems

In the below example, we talk about how blockchain can be implemented in programming. A common problem in database programming is the ability to access elements of it through a database id. It is easy for novice programmers to implement this in a way that is confusing and inefficient; making it difficult to access elements of the database. Some programmers make it too easily accessible; and hackers can easily figure out how to get the database ids. Using blockchain in this application eliminates the need for an id; and fixes both issues of privacy and efficiency at the same time. It uses an automatically generated encryption code or hash key; which is hard to hack. Essentially, the problem of the identification method used in today's databases could be done away with if the correct method of blockchain was used to store user blocks.

Instant Secure Cross Border Payments

A lot of countries deal with the problem of inter-country trade restrictions. They do this because they can monitor the money and activity associated with financial accounts in their own country. They can easily block money transfers to certain countries.

Blockchain has the ability to neutralize the government's control over who you can and cannot have transactions with. For instance, let's say that you want to trade to someone in Russia. You live in some country that is against Russia. You can then connect a blockchain to someone who is able to trade with Russia, effectively creating a three-way pattern that allows you to stay compliant with the government yet do the transaction you want to do.

Secure Identity like IOTA

All Companies Will Participate in the Internet of Things

The Internet of things used to be a concept that was laughed at a very short while ago because no one really believed that it would be happening. I think the time that I started believing that it would happen was when my toaster was connected to the internet. I could not understand for the life of me why you would need a toaster connected to the internet. The fact of the matter is that all companies will fall to the internet of things because in business, it actually makes sense. If you want to have a good example, the dash buttons that used to exist, and still do, are not really needed. But they bring in so much extra income to Amazon that it was a brilliant move.

There's just one tiny problem with connecting appliances to the internet, which is that of security because appliances have not needed regular updates for a very long time. For instance, let's say that you bought a new fridge. It has a tablet-like screen that allows you to keep track of your food, letting you know which foods are going out of stock without you forgetting where they're located in your fridge.

This would be a service that allowed you to automatically add that food type to a shopping cart that would either be sent out to you or you could go pick it up, but it would be ready for you in either case. It would be paid for and everything would have been taken care of. To me, this sounds like a fantastic invention. But the problem is that there's nothing really protecting that application from hackers. You could try to make it so that the tablet was simply a web page, but this would actually have a few of its own complications. These complications include resetting the screen every time the fridge was unplugged by the power because web browsers are usually based off of operating system architectures. So you would have to boot the operating system and have an autorun file that opened up the browser that led you to the webpage, but this comes with its own problems.

Letting someone know what type of operating system you use is a very tricky way for social and digital hackers to figure out how to get in that system. If you tell some random person that your computer runs a Windows 7 they're very likely to not really care. However, if you tell that same information to a social hacker that is looking to get information off of that computer, they will then understand that you have a system they can plug a USB into and get into automatically. They will then understand your computer no longer receives its security updates and that if they really wanted to get into your computer all they would need is your email address. Then they would send you a file that you unwittingly click on while a Powershell prompt comes up in the background, which gives them backdoor access before closing.

Understanding this concept is actually key to understanding why the Internet of Things devices need to have better security around them, ones that are not as susceptible to short range attacks and interface attacks like our modern desktops are. Operating systems usually don't have to think about the attacker directly communicating with the operating system because there are APIs that the operator has to contact in order to get access to the operating system, but oftentimes these Internet of Things devices utilize Linux as it reduces the barrier needed to have control over the device. That isn't to say that Linux is bad by default, but most of them are using very lightweight operating systems so that they can minimize the amount of device resources each device needs. This means that you lose out on security as a result. Linux is an incredibly secure operating system, but these smaller systems gut out a lot of what makes Linux Linux. Credit card information could be transferred over the internet and easily swiped if an attacker found a way to force the device to switch to their internet, which is more common than you think. I won't exactly tell you how it's done but I can tell you that the usual steps are to block your internet first before starting up theirs. Something like IOTA actually provides you with security and encryption protocols that are very lightweight and can be easily shared with other devices just like it.

It's Not A Blockchain But A Blocknet

The reason why IOTA can do this is because it's not really a blockchain more like a blocknet and that is to say that information is contained within the net of blocks rather than a linear chain. This actually allows quite a few benefits to it because the blockchain is centralized but decentralized. Everyone has access to the same ledger but everyone needs to approve a transaction to the ledger in order for the ledger to be changed. This sounds a lot like a blockchain but in this instance, it's actually quite different.

With this type of technology, there is a central ledger that does not have direct access to it. Instead, you make a transaction API call to submit a transaction and this transaction is spread to every node in the network for consensus approval. This would prevent things like altering information or using a credit card where items on the credit card are not 100% accurate and similar fraud charges. All of the information going to the API is encrypted so at every user point the information is already encrypted and the API call sends out a consensus request to all devices that should respond back as soon as they get it whether they concede or refuse. Based on this, the transaction is made in the ledger and then another service can now access the blockchain or rather ledger to confirm that the information has changed and then take certain actions upon it. Therefore, it is a combination of both decentralized and centralized control for the best of two worlds because no one actually directly interacts with the ledger so no one can really tamper with it. Meanwhile, in order to make changes on the network, everyone has to agree with those changes.

Everyone Involved is Involved

There are two unique properties here. The first property is that this type of blockchain is actually very reminiscent of the Internet of Things idea and that is that all the devices are connected in a household, but this idea expands upon it saying that all devices on the internet of things are connected to each other.

The reason why I called this the blocknet is because it's a lot like the internet. Believe it or not, HTML is a ledger of rules that every browser either agrees or disagrees to commit to. Regardless of which device you go to, most browsers will either use HTML, xHTML, and usually that's it. There are a few smaller ones out there but usually, those two are the majority of the internet with some very useful tools in the background. In order to make changes to HTML, there has to be a consensus amongst developers that changes need to be made and reasons have to be given. No one can just willy-nilly make changes to HTML, you actually have to go through a very lengthy process by contacting nearly everyone who's in charge of HTML to see if a change can be implemented and they have to agree with you in order to even do anything about it. Yet, even though it is one ledger of digital rules, thousands of computers follow those rules and update when that ledger updates. There are some unique differences between the browsers, but it's usually quite consistent that browsers stay updated with the latest HTML trends except for whatever Microsoft makes.

The second property is that the blockchain is malleable and can be changed. Sometimes, you might need more or less information. With this type of network, you can section off what type of transactions your information needs to make, which allows multiple devices to interact with each other based on needs. If you recall, most of what happens on previous blockchains was immutable, which meant that it couldn't be changed. With a set of rules that just allows for a transaction of information to occur, you can modify the amount of information in that transaction. This IOTA is very similar to how JSON works to communicate information from the front-end to the back-end in web browsers, which is how *most* interactions happen on the internet.

Enforced Accountability

All Actions Can be Recorded and Non-Editable

A great feature that's part of QuickBooks is the fact that you cannot make edits to information without those modifications being recorded. This is because when you make those audits or editing modifications, you are essentially changing tax records and so all modifications are recorded for the purpose of proving something happened at some date. This makes QuickBooks a lot better than Excel for taxes.

The problem is that there's not much software outside of the tax world that generally holds people accountable. The closest thing that we have to it that is very common in society, is the customer support ticket. With a customer support ticket, the support ticket is assigned to a team and then someone from the team begins to interact with the customer and everything is recorded to that customer support ticket. This provides a way for the company to go back and review the material as well as hold those that dealt with the customer in the customer support ticket accountable for their own actions. The reason why companies want to do this is because sometimes customer representatives have a habit of not doing what the company wants the employee to do. For instance, a customer representative might tell a customer that the customer's problem is not worth their time and that the customer needs to be more intelligent about their decisions. Almost no company on the planet wants this. So, what happens is that the customer makes a review that's bad and at some point in time in the future, the company will either go through a standard routine check of customer tickets. Or the PR around the issue will get so bad that the company will find out and they'll be able to track down the individual.

However, many companies don't have a system like this for anything other than the customer representatives because it's difficult to conceptualize not only what needs to be done but how it is going to be implemented. The blockchain represents the implementation of this problem because with a blockchain you are able to ensure records are being kept of people's actions and you can begin implementing it in nearly everything because you already have an implementation standard. This type of blockchain allows for companies to hold themselves and others accountable in the market of ideas and in the court of law.

Blocks Can Be Assigned to Members

The best part is that both the customer and the employee can stay anonymous until it is absolutely necessary to require their names. The most common way to figure out a system is to look for developers that work within the system and then pay attention to the internet when that person seeks help on forms. A great majority of a developer's initial work is built on getting help from others who are more knowledgeable than they are because it is impossible to memorize everything about programming.

The problem with this is that when you're dealing with customer support tickets that have to do with the platform, you're looking at being able to attach a real-life name to the person who both is asking the question and the person giving me the answer. Social hackers understand that certain bugs are fixed in different ways depending on the programming language you have. A huge problem in figuring out the weaknesses of a system involves figuring out what language that system uses because even programming languages are subject to bugs and problems. Therefore having real names attached to it causes of a huge issue because when that developer begins to ask questions about the back-end language, they are going to use the code that they're having a problem with and this lets the social hacker understand what programming language is used for the back end, which makes the back end even more vulnerable.

I have said in the past, blocks in your average blockchain are primarily made up of a timestamp, a worker ID, and a coin ID. However, in a valuable blockchain system you can actually have it so that instead of having a worker ID you have a random digit that represents an employee and a random digit that represents a customer and have two different databases for them based off of a specialized ID. This hides both databases and the names associated with customer support, hides the name of the person submitting a question, encrypts all of the information, and creates a system of accountability.

A New Supplier Network

This new system of accountability can actually be used to create a new system of supplier networks because a great portion of the work in a supplier network is verifying that something has arrived, and something has left. It is very difficult to build a system that will stand the test of time that ensures that something will arrive, and something will depart. For example, let's say that we own a packing company the allows you to send various parts out. Normally, the person who makes the part can only ever put a modification on the part and this is a problem because the part manufacturer has to rely on the carrier to verify that something has been delivered. This can lead to lost packages and frustration in some cases.

In the Internet of Things, most things are online. The best part about using a blockchain is that you can attach a GPS to it by putting a GPS unit in your box that will transmit the location of the device. It is also transmitted in a format that is very secure. You don't want to willy-nilly give someone else's address away. This is the primary reason why part manufacturers do not put a sticker on the side of the box is because it doesn't transmit the data and it doesn't allow for secure information to be shared over the internet.

Recruitment Reference Checks BEGONE!

As I've mentioned many times in the past, the primary problem that we deal with in middlemen is that most of it is dedicated to simply verifying if the information is true. The reason why this is difficult to do is that you have to have systems that people can trust before people will trust that information. Naturally, if you have a driver's license then the most common thing that the person will trust is the maker of that driver's license. So what you really have is a government that makes the driver's license and the place you're using the driver's license with wants to check with the government to see if it's legitimate. This usually involves a very small fee for the work that's involved with checking that driver's license.

One of the problems that could significantly be a benefit to people utilizing the blockchain is the idea of secure references. A big problem with references is that you tend to have to trust people on both ends of the system whether it be the person who's applying for the job or the person that was actually listed as a reference, both of which could be lying. This is an enormous problem for security positions in areas where you generally want to make sure that neither of these two people are lying but you have situations where it's a possibility and it's actually quite difficult to catch them in that lie. This is until you incorporate blockchain technology into references because blockchains can be immutable.

Let's say that you have been working with a company for a number of years and when you first started working with the company, part of the process was submitting your name to a reference company. Essentially, this would be an employer reference found on a resume but the information about whether you worked for a company or not would be stored by a company that provides this as a service to many companies around the world. This provides verifiable proof with minimal fees that you were part of an organization because the company has to have direct access to the section of their references. This allows you to do multiple things.

First, you can have a company sets up a reference database for everyone that works at the company so that other companies can figure out the people who are lying and the people who are telling the truth. This reduces a significant amount of the work involved in running background checks and ensuring the equality of work that's going to come from a specific employee. The employee ID number can be used to provide information about the employee work history without actually having to provide confidential information (this differs based on local law).

The last thing that can be done with this is actually something that involves security for those who apply to jobs. The primary problem with resume websites is that unless the name is very well known in the PR space, you don't really know if the resume website is legitimate or not. For instance, a common scam that's going around at the point of writing this book is a job offering from a company that disguises itself as Amazon and then links you to something like AmazonInfo.org and this means that when you submit your information to this website, you now have to deal with the repercussions of being scammed. This scam website wants your credit card information to set you up with a kit that will "prepare you to work at Amazon" and then once they get your credit card information they also have your resume which usually includes your phone number and your email address so they can begin blasting you with scam calls and scam emails. This is a huge problem and there is no real way to stop it other than removing the individual and that individual will have another individual just like it pop up very shortly after or during the stopping of that individual.

Something else that you can do is have a verification of resume websites so that not only employers have trusted websites they can hire people from, but they can also provide security for the potential employees looking to obtain a job there. Instead of relying on websites that tell you that they are safe or legitimate, you can go to the well-known reference company that the website uses to verify that they are an active user with that system of references. This allows security for employers and employees while also making the life of employers and employees significantly easier.

Payments with No Middle Man

Cryptocurrency Has No Border, Et tu Blockchain

The benefit of blockchain also comes in the form of how you can connect people together because a lot of the verification systems and payment systems used today require government cooperation. The government might want to sanction a certain country based off of their actions. However, you as a company have significant dealings with that another company in that country. This causes such sanctions to practically bankrupt your company. This is not much of an issue with a lot of home-based companies that don't work internationally, but companies such as microchip manufacturers and things like that might find themselves in a pickle when the majority of microchips are made in China. Yet, your government has decided to block most of the trade coming from China. Essentially, creating a huge scarcity that's purely political when there doesn't need to be one in your company's vision.

A blockchain doesn't require government cooperation as of right now which means that you can essentially have a system that ignores the need for government if you absolutely need to. Mind you, there are border customs and different situations when you are shipping things like physical products but in the case of something like software keys or schematics, these are purely digital and most of the problem deals with having digital work done but not being able to pay for that work or being taxed on the pay for that work at a significant rate because of this political issue.

With a blockchain, you can register a customer in your accounts that stays permanent and you can apply cryptocurrency to that customer. The reason why this is important is because by being able to tie cryptocurrency, a currency not currently controlled by the vast majority of governments in the world, you are able to avoid being taxed on paying or being prevented to pay for that work. You can essentially convert your money into cryptocurrency, exchange it to the person who did the work, then that person can take the cryptocurrency to an exchange market and get the money not only from the cryptocurrency but in the currency that they prefer. There are actually systems like this in place right now, but they are really small and a lot of people don't trust them because they're usually not made within the country that they live in, therefore they are very suspect. Also, a few countries have banned the use of cryptocurrencies to prevent this; so this could be an issue as well.

In-House Cryptocurrency

This is a very unique idea because it's not really one that surfaced in the middle or even mainstream market and is still relatively small. An in-house cryptocurrency allows you to pay for things at companies using your allowance of cryptocurrency that the company provides you with. If you think of companies that incentivize work that's done fast or work that's done to a very high quality, you can think of the same system where kids are given coins for doing good jobs at their school and being able to amass a certain amount of those coins to earn prizes. This is very similar to that because the reward system is actually in use in quite a few countries except for places like The American or The European system. The reason why the reward system exists is it be encouraging workers to do more than just the bare minimum.

Workers understand that they will get paid if they do their job but many workers, when they come to realize this, tend to only do the bare minimum amount of work at their place of work because they know they'll get paid. When someone knows that there is no incentive to go above and beyond what they're actually being paid for, it is very difficult to get cooperation within the workplace and so things slow down significantly. In America, the way that they like to reward good work is through parties, company get-togethers, retreats, and every once in a while you might actually hear about a company that rewards you with a holiday present ranging anywhere from turkey to something like a cruise that you may have wanted to go on. The reason why these companies do this is because it creates a personal connection with the worker of that company but the problem with personal connections is that it causes personal problems (Of course, we are not talking about big banks here; which is a completely different story).

If a person works for a company for a decade and they have begun to think of themselves as part of the family, they will indeed work more than they need to so that everybody in the family stays happy. No one in the family likes the person who does the least amount of work because there's always something to complain about and so more and more reasons are created to get rid of that person. It creates a sense of competitiveness but in a less direct way. Companies that reward their workers with things like a new iPhone or Samsung do so based off of company goals that are directly increasing the

236

competitiveness amongst their workers. The difference between the two is that workers that are compensated with gatherings and things that allowed them to enjoy something with other workers creates a Cooperative environment. Workers that are incentivized to do better because they will earn prizes actually creates a cutthroat environment. You see this in police dramas and shows depicting office reporters where they often try to outdo one another so that they can get more benefits than the others in the group. Those that ultimately do get those benefits are actually despised by everyone else and so it eventually evolves from being a collective group of cutthroat individuals looking for benefits into a strong brotherhood of rotational benefits. This is highly detrimental and is definitely very centered around who manages to get what.

In-house cryptocurrency is a different option than both of these because cryptocurrency by itself actually promotes the ability to exchange for money or for items. Therefore, you can think of it like a mixture between you can either choose to get a bonus paycheck or you can win the cruise prize that you might have wanted. The best part about this type of In-House cryptocurrency is the fact that different amounts of the cryptocurrency can be distributed to all workers at the same time. This also provides a level of performance checking to see who's doing the best and who's doing the worst. It's a very unique combination of performance checks, reward checks, and co-operation encouragement. You see it encourages cooperation because everyone gets rewards but the person with the lowest amount of rewards is the ultimate denominator of how much everyone gets, and it creates a weak link that encourages others to help that weak link.

Ripple: Earthport Service Payment System

What is it? It's Not a Blockchain

Ripple is a very unique type of cryptocurrency being used to provide an easier method of transactions between banks. It is something that can be invested in, but the primary use for it is for banks looking to do cross-border transactions where it would normally take a business week or longer. It was actually developed in 2012 and is recently a hot market because of the enormous value boost that occurred in 2017. Ripple is a Protocol and not a Blockchain, which means it's a set of rules for banks to use in order to make transfers *through* Ripple Network.

XRP Tokens Finality with Drops and IOU Non-finality

XRP is the name of the currency and is primarily what is used to make transactions. Ripple also makes use of something called an IOU, which is a promise that a transaction will occur eventually. IOUs are agreements between banks where one bank does not have yet enough money to make their half of the transaction, but the other bank does and the other bank agrees to the transaction, accepting the IOU that can be redeem. This creates a loan-type system that banks can use to continue making transactions, which allows the market of business to continue.

XRP, unlike IOU, has a finality to it. There are exactly 100 Billion XRP, but it's important to remember that cryptocurrency coins can be spent in the billionths or more so a transaction might be 0.0000000000000000000000000012 XRP fee, which means it's not a big issue at all.

Marketmakers: Buy and Sell Orders

The way that trades happen are through devices/people known as Marketmakers. When someone submits a request, they submit it through XRP by exchanging something like USD to a Marketmaker that will buy it with XRP. Someone else submits something like CAD to the same Marketmaker that will buy it with XRP. That Marketmaker will then buy back the USD person's XRP with CAD and then use the USD to buy back the CAD person's XRP. At some point during this transaction cycle, the Marketmaker will charge a fee of their own to make money off of the transaction. This is how Marketmakers can make money and how the Ripple Net functions for the most part for cross-border exchanges.

How BlockChain Technology is Implemented in Smart Contracts

Ethereum and Smart Contracts

There are 2 Ethereums out there, Ethereum and Ethereum Classic, which is a somewhat important thing to point out. They both have Smart Contracts in them, but they are two different platforms. The reason why the difference is important has to deal with Decentralized Autonomous Organization or the DAO.

The DAO would serve as a Networked Hedge Fund, something investors could buy into if they wanted to support some organization through Ethereum. Investors would by DAO tokens and give them to Decentralized Applications or DAPPs, which allowed them to invest and get voting power like stocks do. In order to buy those tokens, investors would have to use Ether, the transactional money of Ethereum.

The problem was when investors wanted to leave this little ecosystem, they needed to use the "Split Function". This function served as a way to exit the ecosystem, get the Ether they invested back, and gave them an option to create a Child DAO that would become an Ether Investment Firm of sorts. With intelligent inventions often comes intelligent exploiters, which meant this system of leaving was exploitable as the DAO had to be computed. A hacker made a recursive Split Function, which caused a continuous release of Ether until ⅓ of DAO funds were released into their pockets.

This massive issue stole a lot of Ether, affecting both DAO and ETC (Ethereum Classic) communities until the ultimate decision was made to split those communities. ETH (Ethereum) serves as a new cryptocurrency with a blockchain that prevents this while ETC allowed those who had amassed Ether to continue operating. With a new coin on the market, Ethereum was created and the community that had left managed to get the funds back.

With Ethereum Came Smart Contracts

ETC became just another cryptocurrency while ETH became a cryptosoftware company with a mutable blockchain. While the fast majority of the community feels that the blockchain is immutable, it can be changed by those who hold power over it and this was needed in order to get the money back. By making it mutable, it also made it open to certain changes like the ones we're talking about now.

ETH, unlike ETC, is backed by what's known as the Ethereum Alliance which is comprised of some very high value firms like Microsoft and JP Morgan. This means they have a lot of weight that can be thrown around, and ETH is now worth almost 15x ETC.

What is a Smart Contract?

Proof of Stake

As mentioned before, cryptocurrency makers wanted to do a better job at making less of an impact on computer power so now there's something called Proof of Stake. The way that this process works is that instead of creating a difficult problem for people to solve, rewards are handed out to random people on the network based on the stake they have in the network.

The reason why this switch took place is that it is far less prone to a takeover and less resources are used to maintain the network. Essentially, all you have to have in order to make Ether in Ethereum is Ether itself. The more of it you have, the more likely you are to make money off of it. Bitcoin had a takeover with mining farms, which could horde all of the earning because a single computer couldn't compete with something like that. As a result, the cryptocurrency has an active market where everyone tries to get a bigger stake in the network.

Additionally, if you tried to tamper with the network or the transaction then you risk having all of your Ether taken away from you or the reward. This would be automatically decided by the Blockchain when it detected tampering.

It's Still a Blockchain Sort Of

It's important to understand that a Smart Contract is still a coin, it's just that the coin has the contractual state inside of it. Smart Contracts are rather interesting because it is a saved Code State Listener. To understand what a Code State Listener is, you have to understand the difference between Parallel and Linear Processing. In computing, everything happens in a linear processing pattern, which means everything goes from Point A to Point B.

Having said that, Parallel Processing is where Linear Processing happens at the same time so while you go from Point A to Point B, someone else goes from Point C to Point D. Now, a Code State Listener is when a Third Person is waiting on both of you to arrive at the same time in order to begin. This is how Parallel Computation is done.

In a Smart Contract, there's a bit of code that waits for Conditions to be met before running a completion or at-end-of-program code block. Therefore, a Smart Contract waits until you do what you need to before it begins running.

The way the blockchain works in this system is that *it* is the one that enforces that the contract is followed. Therefore, someone creates a Smart Contract, that Smart Contract "listens" for your conditions to be met, and when those conditions are met the blockchain executes the Smart Contract.

Smart Contracts Are Not Applicable to Everything

Conditionals Are Binary

One of the problems with a Smart Contract is that it needs to be an either-or choice, it cannot be multi-conditional. For instance, let's say that you have the option to rent a car or buy a car. If you rent the car, the Smart Contract has been fulfilled according to the blockchain and no more will be demanded. Essentially, they will have to have a new Smart Contract for months on end until the company sees it as paid.

Smart Contracts cannot have Recurring Events, which means something like renting requires new contracts with new signatures.

Conditionals Have to Be Binary

Conditions also have to be quantifiable in a "Did" or "Did not" manner, which means partial pay is not an option. Let's say that a consumer wants to pay for a specific Service and that Service is tied to a Smart Contract. The Service may cost 300ETH and thus, unless 300ETH is paid to a specific account, the contract is still live. However, the consumer offers to pay for a product that your team needs like a new editing software and wants to compensate the cost by modifying the contract. This cannot be done and while there are ways around it, it still represents a limitation of Smart Contracts.

Listener Overload

This has not happened yet, and I am surprised that no one has figured this out, but there's a potential for a listener overload. Each time a Smart Contract is made, it has to listen and that means code is running in the background to make this possible. Thus, you can recursively open listeners with small contract loads and you would need a *ton* of ETH, but it is a vulnerability of the system. What I think would happen is that the entire system would become incredibly slow to the point where some might find it unusable. Beyond these things, Smart Contracts are applicable to a lot of situations and that makes them incredibly useful.

The Future of BlockChain in Marketing and Sales

Universal ID Distribution

Non-User Based Tracking

Most of the issue is often about gaining new customers and, as we've seen with the Mark Zuckerburg conference, a lot of people don't like thinking that their personal information is tracked when they're not a user and didn't sign anything. The beauty about this is that Blockchains can ensure their information is secured and they can be assigned a Blockchain number instead of trying to identify who the user is.

A lot of personal information goes into "shadow profiles" and many of them simply record the type of websites you go to because keylogging is illegal. They also record your mouse as you visit their website so as to see if their designs are luring people in. This is data that is ultimately disposed of at the end of the month as it is difficult to maintain storage around this, but those that are not tech-savvy don't have a method of being sure. However, if those people understand that their information is held in a blockchain, is completely anonymized, and the entire blockchain is thrown away and recreated, such fears might go away.

Success and Fail Profiles

Attached to the concept of anonymization are Success and Fail profiles, which refers to when a user clicks on the call-to-action or they do not. The sad thing is that most websites have to tie this to some identifying marker, with the easiest one being the IP address that comes in on the marker. However, if everyone shares data then this becomes an issue of targeting people at their devices, the primary reason why VPNs have shot up in use. Knowing this, the Blockchain produces a very interesting opportunity because recording IP addresses is no longer viable at this point, but a Blockchain ID is still usable. Essentially, it'd be nothing more than a Javascript cookie that collects information, much like what is used today but this information would be based on randomized ID rather than the IP address that connects to the server. Then you could create anonymous associations like the language used or the currency used, which are detectable default settings. It provides a far more targeted way of tracking without actually know who the person is based on a cookie that you dropped in. This allows you to create these profiles to a much more accurate degree without sacrificing privacy.

Advertisement Auditing

Showed Versus Clicked

One of the most expensive things in advertising is just proving that the advertisement has actually been effective in causing an action from the user. With all online advertisement methods, there comes a point where the advertisement either results in a click or not a click. This would seem like a relatively uncomplicated problem, but the problem comes in the form of companies that get paid by those advertisements.

Let's look at the Google advertisement program because this is a program where you can obtain a key associated with your account that allows you to get money for advertising inside of your product. Websites and applications both use this form of advertisement interaction because it's really easy to implement but it is really easy to fake as well. One of the most notorious problems with online advertisements is the determination of real clicks vs. fake clicks because fake clicks are caused by companies who want to earn money from ad revenue without having a lot of customers. Essentially, the old way of doing this was to set up a web browser on a computer and continuously reload that computer and click on an advertisement only to shortly thereafter close the link that the click led to. Now, this is a lot less common than it used to be because it's quite easy to figure out whether an advertisement is being clicked on because the company wants to make money from it or the customer is actually engaged with the advertisement itself.

The way that companies stopped, mostly, fake clicks in this form was to tie the click to an address but this just elevates the problem to a more complex problem. You see, you have things like VPNs that allow users to change IP addresses so that even though you detected a click happened just five seconds earlier, it looks like it's coming from a different address the next time that the click happens. The reason why this is difficult to detect is because you have to either assume that it is a different click, or you have to find the MAC address to the device that's doing the clicking.

Web browsers have naturally been able to implement this, and the fake clicks are usually people who get paid to click, but that's also really difficult to track. Facebook is notorious at this point for having these fake clicks because you might see thousands of people clicking on an advertisement but the only way that you know that they're fake is if all the clicking comes from a country that has a low GDP. Even then, it's speculation and not fact at that point because you can't exactly prove that it's a fake click yet plenty of people do this. Again, companies have tried to combat this by making it illegal inside of the terms of service to do such a thing but it's really difficult to track.

However, with the blockchain you can track it. With the blockchain you can tie known addresses to previous users in a very small format. You can actually do this with users now in a basic spreadsheet, but the complexity comes from how you go about verifying fake vs. real clicks. For instance, a good majority of how fake clicks are detected today are usually associated with the behaviors of the person doing the clicking and this is why Google gets paid a lot for their advertisements.

With a blockchain, you can access several points of data that might normally have been off-limits to you before. For example, in the regular world you do not have immediate access to the GDP of a country at the same time you have access to the MAC address of the computer that's clicking on it in the same spreadsheet but with a blockchain you can tie the current GDP with the current MAC address and begin following that Mac address to see what the computer does as it relates to your advertisements and how often those clicks are made.

Further information can be given at that point such as how much that person makes according to their reported amount on another blockchain, where that person lives, perhaps what that person does on an Android phone, and the information goes further and further and further because people trust security-based Systems. Therefore, things that have previously been off-limits such as a person's address as it relates to the advertisement can now be made available to the Advertiser because that Advertiser is a trusted member that isn't going to use the address for bad things. Because you cannot tamper with the information, blockchains provide a way of accessing encrypted information, because those that access the blockchain have to be pre-approved, and because companies tend to work together, you have a situation where you can now track the behavior patterns and verify the existence of people in countries that you would normally not have those rights in.

Having 100 cell phones connected to the same IP address with all different Mac addresses can be simulated but knowing that they're connected to the same IP address and knowing that that IP address has been consistently flagged in the blockchain as used for fake purposes creates a verified way of ensuring the difference between fake clicks and real clicks. Such a process saves millions of advertising dollars.

Company and Product Authenticity

Anyone Can Use A Bar Code

The truth of the matter is that pretty much anybody that has a printer can create barcodes of their own. This is true of trademarks and logos but with the blockchain, it becomes a different ideal altogether. It's been a very long time since the standard barcode has been updated and the problem is that these markers on these boxes are used to identify organizations. For instance, you will commonly see the FCC logo and the FTC logo on electronical products depending on their uses. You will often see certification logos on the sides of desktops to certify that the product is working as expected. However, these logos can be duplicated because all you need to have is an image of that logo.

Photoshop is a fantastic tool but there's one problem and that it is able to create anything from a photo. This means that Chinese (which is a country that's most notorious for this) manufacturers that are not legitimate can take the FCC Logo and put it on the back of any technology that would require it. This allows them to trick users into believing that the item has been checked by the FCC and it is an aspect of the used product market that many IT professionals have spoken about. This brings us to the next topic though.

Branding With Special Font Encryption

When you deal with barcodes, you are creating an item for scanning and the newest form of barcode is actually known as a QR code. Those little weird boxes that provide additional information with them. You need a QR reader in order to translate them, but they do give extra material on top of what would normally classified as a barcode. With blockchain technology, this becomes a different story because now you can have a logo that has unique ink that only shows up in a phone. All you need to do is scan with your phone and this can legitimize a product or delegitimize it. Essentially, you can have as much encryption as a dollar bill on your own branding that a customer can then scan to access information on a secure server that verifies that this is a branded product.

The reason why this hasn't been doable in the past is because smartphones have notoriously not had the best of cameras, there was no way to secure the code in a way that copyright infringers couldn't copy, and there wasn't a way to secure that information so that it couldn't be tampered with. That last one is very important because when you let people know where the information is, bad actors will try to find out how to get access to that information. If the information is part of a blockchain, you can now modulate access to different verification blocks. Therefore, only your company can transverse the blockchain itself while the customer sees a block in the chain, effectively isolating it from the rest of the chain. Plus, the scan material itself can be an identification marker such as a certain combo of red and green might pick up Starbucks or Chugg, which would then cause a lookup in their database that no one else can access.

Products Can Tell A Story

One of the unique traits of an item like this is that each product can tell a story. Unlike the Barcode or the QR code, a lot more information can be tied to a block. This allows a story of the product to be contained within the scan, such as where the materials came from, what went into the product, and a lot more.

This creates a personal connection between the company and the potential consumer.

Anonymized Marketing

Personal Ads Without Being Personal

Since the advertisement information isn't personal information, one no longer needs to risk an incident like the "gay caught at work". Instead, each advertisement is tied to a blockchain profile and that profile can then have safe and unsafe advertisements. Therefore, if Lucy the lesbian opens up her browser at work, this identification number is not the one she has at home and so only safe advertisements show up at work. Additionally, work environments can specify what type of advertisements they want through this system, such as adverts targeting based on sex appeal or "LGBT" items, which can cause controversy amongst those that pass by and those that wanted to keep that information private.

Ads Fit the Audience and Sometimes the Person

In addition to this, adverts that are hosted through a blockchain can meet audience needs rather than an individual person. One of the most annoying things about adverts is when you get them for totally unrelated items. For instance, one video that might be talking about HTTP 2.0 has an advert off to the side that's talking about the new Furby or this excellent gaming website.

Based on Blockchain information, this blockchain id can be tied to what adverts were successful and which ones were not. So, if a tech advert is more successful for this person, Furby electronics and a game won't show up. However, if that person is interested in tech but has shown to have clicked on game adverts then the game advert could show up but the Furby electronic would never come up. This speaks to the tech audience, not to the person who might be interested in games. The difference is that a tech audience is far more likely to click on an advert for server tech than they are for a game unless that specific person has done so in the past, this increases the likelihood of successful adverts.

Blockchain Profiles and Saving Money

Make Money On "Adcoins"

Adverts are not something people want to see, pretty much ever. Knowing this, a company that wanted to support creators came out that was known as Brave. The way that Brave works is that a user can put money towards people they support through micropayments. Most of the reason as to why we see advertisements is because the website wants to make money, so users help support them by themselves. If 1 million people watch an ad-free video and each of them gives over $0.01 USD, the creator can get $10,000. To give you an understanding of the difference, usually 1,000 views = $1 and so putting 1,000,000 to that you get $1,000. So, in this situation, the Brave app actually pays more, which means that adverts could be avoided altogether and make the creator even more money. Secure transactions can be handled through blockchain as Brave suggests that they do.

This system is brilliant, except for 1 issue and that is that it removes advertisements completely. Brave *also* handled this because it has a Rewards Program for those that do not mind watching adverts if it means they can be rewarded directly for doing so, which is then something that can be given to support creators. Essentially, Brave created an ecosystem that allows users to prevent websites from tracking them, rewards them for adverts they don't mind watching, blocks annoying ads, saves data, exchanges in cryptocurrency, and provides a new way of searching the internet.

Their system makes sense, but the truth is that other browsers should also be complying with this and they're not. Users are constantly submitted to watching adverts to the point that items like Adblock Pro are used just to ensure an ad-free experience. Creators hate advertisements as much as the next person, but people would rather support their favorite creators directly if they could. Possibly the worst part about the internet is the invasive ad system unless you incentivize users to it. The website gets paid to host the advertisement, users should be paid to watch it because they *don't* want to watch it.

Companies have a hard time conceptualizing why this should be the way that it is because companies think they'd have to spend more money. People know websites only use advertisements because it supports them as a website. Advertisers waste a *ton* of money on ad campaigns that target entire classes of people who could care less about the product. Even worse, most websites get paid a fraction of a penny for a view and maybe a penny for a click. Therefore, you have a system where a ton of money is being wasted because you don't know who you're advertising to *exactly* and the only information you're relying on is the tracking history of web searches and desktop applications. On the other side, websites make a decent living if they can pump out enough content so that you'll look at adverts enough times to get paid and hopefully people will sometimes click on ads, which means it's a costly wasteful gotcha system.

With a system like Brave, websites are supported directly by the userbase and often much more so because users don't know how much to give so they usually over give. Those that don't have money opt-in to an ad program that will give them money that they can then contribute using a no-border cryptocurrency that can be sold in an exchange for money. Those advertising in the ad program have access to the likes and interests of everyone they're adverting to, so they know exactly the customer they are selling to and they only ever spend as much money as they want versus a situation where they have to estimate a budget. Like Facebook gets $50 for 5,000 views with no promises of clicks or even how targeted it truly is, whereas something like Brave allows you to pay out money for every full viewing you get to someone that's extremely well-targeted.

Possible Government Compliant Ad Database

Finally, this system is non-invasive and allows users to participate in a way where there's no database of personal information just for advertisements. Advertisers don't get access to location, places visited, or other information that even the government might be jealous of because **they don't need it** as they're more interested in what type of customer, they're selling to than where they live.

This means the possibility of a complete Government Compliant Ad Database can be made, where companies can submit advertisements that can rated. Kids don't need to see porn and adults don't need to see "kiddy games", which creates an ESRB-type system for advertisements. What information can be collected can finally be regulated *by the government* and no company that's built on selling invasive information to advertisements would have to switch to a less invasive model. It would lower data costs vastly for the majority of people, protect people, de-incentivize NSA-level spying, and increase the effectiveness of advertisements while reducing targeting people individually.

No Leakable Points… Sort Of

Perhaps the best part about this sort of blockchain technology is that there are no "leakable" points. You no longer have to collect Big Data via tracking just to get the edge on marketing because people will hand you their data! Such a system opens up an entirely new world of advertising, one that focuses on getting feedback from customers without knowing who the customers are. A survey can be asked to determine company image without it being shoved inside of people's content. You can have several different ad campaigns to see which ad does best with which crowd for far less money.

This means that all of this information is passed over a secured connection to a blockchain that's encrypted and only accessible to people who are authorized, by the consumer, to have access. No more guessing, no more useless profiles, no more wasted money and advertisements mean something again.

How BlockChain Technology Improves Online Safety

Secure Storage and Complete Encryption

Encryption with a Unique Algorithm

The beauty of encryption is that it can be carried out in a number of ways, which means that security is actually a factorial problem. So, let's say that you use SHA-256 to encrypt your stuff, which is insecure, and you shouldn't, but it's one way in which you can encrypt items, which is important. Had the method not found a way to be broken, it would take less than a day to break it.

However, what if you encrypted it with MD5 (still somewhat good)? Well, this would take about 6 months after the hacker *figured out* that they were dealing with an MD5 encryption and then take 1 day on top of that to break SHA-256. That's pretty good, but can we do better? Yes we can! We can mix MD5 with SHA-256! At this point, you're looking at a ridiculously long number that would cause you to die along with 300 generations before you ever got to the point of breaking what's there. Believe it or not, that's the type of encryption that's on the blockchain.

Obscuring Data Location

Half the battle in finding sensitive information is hiding information, because when a hacker comes into your system that's what they're looking for. The way that hackers find information are through bugs, security holes, and people. This means that if you want to hide something like a blockchain, all you have to do is create as many barriers to it as you can. With a blockchain, this is quite easy because every block in the blockchain is going to be impossible to find unless you know the exact identification marker used when finding it. If you can figure this out, you are halfway to figure out how to find the others. However, once you do you then need to decrypt it and unless you know the salt (added in word to prevent repetitive patterns in encryption), good luck every doing that.

A Backups Backup

The interesting thing about a blockchain is that it can be used to backup and verify that data hasn't changed. In a company backup, there's usually a way to keep that backup secure also known as source control. This allows developers to keep a secure copy of the code on hand so that if something happens that went horribly wrong, it's as easy as copy and paste. However, backups tend to go untouched until you get to the next stable point, so they just sit there, vulnerable. A social hacker could get into the building and copy the backup for their own use, which means there needs to be a way to secure it and ensure that everyone has a backup. A Blockchain can be used to do this.

Restoration of Modifications

7 Years on Everyone's Computer

A blockchain can be timestamped so all of those tax records the IRS requires that you have for 7 years can both be automatically store in a blockchain and regularly released every 7th year. This reduces the need for bulky paperwork and ensures the security on those files, but it can also be stored on every secure company computer to give you a way of backing it up everywhere. Not only does this serve as a way of backing up the information so that it isn't lost, but the financial department doesn't need to go searching for files because it'll be on their computer and they can't change it once it is in there.

You Can Backup Blocks Or Blockchains Entirety

As already stated before, you can have different segments of information. You can have blockchains that are contained within blocks, which means you can make an entire database from just blockchain technology, get all of the security you would have from a blockchain, and have an ecosystem built around this.

No De-Platforming

What is De-Platforming?

De-platforming is the act of taking someone's platform away. As of writing this book, Patreon is in trouble with a YouTuber known as Sargon because they evicted him from their platform for off-platform behavior. Currently, people are leaving the website in droves and a few of the top earners are talking about creating their own service, but one thing has come up in all of them: cryptocurrency. Cryptocurrency removes the barrier of entry for many websites that want to exchange money. For instance, Patreon must work with PayPal in order to process transactions, PayPal has to go through Mastercard or Visa, then this goes through the ACH, and the goes back down the tree to finally get to the person that needs the money. That's about 4 gateways with 3 companies that could turn it down. With the blockchain and cryptocurrency, it's from the consumer to the provider. There is no middleman with cryptocurrency, which allows these companies to operate without needing approval from other competing companies.

Blockchain Cannot Be Blocked Directly

The best part about the entire process is that even if a company was told no by another company, they could still find another way to use the blockchain even without cryptocurrency. Cryptocurrency has no borders, which means that if the company can't get approval through PayPal then it can go through some other country's payment processor that's friendly that will then funnel money through the ACH. This effectively destroys any politically motivated company from denying a user money.

Blockchain Has No Borders

Cryptocurrency really has no borders, which means that blockchain has no borders. Imagine a country that was trying to get money for a revolution. Well, if the people had their money in paper form this could be confiscated and lost when an oppressive military invaded. Cryptocurrency is on the internet, which means they could easily save it to some online system far away from it being taken. Essentially, oppressive governments could never de-fund revolutionaries because that country would have to shut down the internet to do so and that can be detrimental to a country's economy.

For example, the economy of Egypt has been significantly impacted by a devaluation of their currency, among other factors ever since there have been significant restrictions on their use of the internet. Sure, there are other factors involved in the Egyptian economy, but the internet restriction is a contributor.

Limitations of the BlockChain

Public Acceptance

People Generally Don't Understand Technology

Perhaps the most frustrating thing about everything is that for this to be useful it has to convince the public. For the majority of history, things are usually built with a use. Oftentimes, that use has to be plastered in front of the user's face several times before they understand the importance of it in their life. That doesn't mean that users are stupid, it just means that people, in general, have a very hard time changing their habits for the needs of the market as the market changes its habits for the needs of the customer.

This means that even though a technology is fantastic and has many applications, it does mean that they are also going to look at the technology like it's got a moral attitude. If trustworthy brands demean the value or associate the use of technology with criminal elements, they are more likely not to use it. Essentially, even though that tool is very useful, if a social association can be made that attaches that technology to something bad, then people are very mistrusting of the technology. You could explain it all day long, tell them about how fantastic it is, but until somebody they like endorses it you have a very small chance of getting them to go on board with it.

Campaigns from Competitors

There's a lot of money to be made by preventing blockchain from being implemented in the market. Remember that the blockchain does away with the middleman and there are entire companies dedicated to being a middleman. These middlemen have created campaigns to make a social association of a negative nature with blockchain to make sure that it doesn't take over the market.

For instance, the wealthiest middlemen are dealerships between the factory and the consumer. If blockchain technology is implemented with selling cars, dealerships will have a huge problem on their hands when consumers can bypass their monopoly-like laws to purchase brand new cars.

Most of what prevents a person from buying a brand-new car online is the fact that it is difficult to get it shipped to your house, it requires a lot of verification, and then it requires a lot of background fees. These companies make money from selling the car at a higher price than what the factory sells to them for and the interest rate on the monthly payments of the average car owner. They often hide in false promises such as if your car breaks down, they'll fix it only for the customer to find out later that they'll fix it if it broke down in a certain way. It's not necessarily a false promise, but it does come across as a false promise because that was not what was told to them but that was what was on paper. People often pay attention to what is sold to them via the voice rather than what's on the paper. To some people, it honestly feels like legal papers are made long and boring, so people sign on them without reading the fine print. That's just a small detail in the grand pattern of and all of these companies will generally get away with making tons of money without the consumer knowing until after they've had a horrible experience. This leaves only the customers who don't really need those extra benefits to promote the company so that more poor individuals have bad experiences.

These people don't like the fact that you can have a solid verification pattern with the blockchain. They need to verify your driver's license, your signature needs to be verified, they need to run a background check on your credit score which requires further identification, and by the end of it they have so much identification they may know more about you than many police officers or even the guy who stole your credit card sometime. All of this can be done online and over the blockchain.

If the government implements a lookup system for driver's license and incorporates some sort of login feature to prevent users outside of the network that are not trusted by the government to get in, any major seller like Amazon or Walmart could take your license number and run it against that database to verify your identity before having a customer support agent come online to record you as you sign a piece of paper saying you bought a car, which is then stored in that same blockchain that they did thousands of other customers along with the contract so that it'll never be lost. All of the information that the sales associate might need could be accessed via a blockchain, requiring no more than digital access. This is a very difficult case to fight for the existence of dealerships beyond being able to look in the car and feel the experience in the car. However, this is also defeated because now you can experience what it's like in a car by putting on a VR headset and sitting in a real-sized version of that car so that you can see what the car is like.

Enormous middlemen brands have begun to notice this and are campaigning to make blockchain look like it's the primary tool for criminals in an attempt to get the technology banned. It doesn't matter if the technology would make people's lives easier, more secure, and overall provide a better experience, it just matters that these companies continue making money without having to make any major changes. Companies do this all the time, with Apple practically being a spokesman about this with constant issues like suing products coming from China simply because they had an Apple logo on it, because it was a reused Apple part, trying to pass it off as fraud.

Cryptocurrency Bane

The sad thing is they don't need to do much work to get this point across. Cryptocurrency had a very large campaign against it by companies that support Fiat money. They would say things like it's criminal money or it's the money used on the dark web, which leads people to believe that it's a bad thing because they are bad people using it. To the average person who has an IQ more than their teenage kid, this is utter nonsense.

However, whether or not it is nonsense, a great majority of people believed it and that is the primary crux of the problem. A lot of people believe that cryptocurrency is a bad thing for a variety of problems. It's associated with criminals, it has an unpredictable stock market, and many other reasons caused people to veer away from cryptocurrency. The reason why this is the crux of the problem is because when people hear blockchain came from cryptocurrency, all those previous associations are now attached to the blockchain. This is a huge limitation for the technology and it works to the benefits of every company that would have automatically been affected by blockchain technology.

No Mainstream No Acceptance

All of this underpins the bigger problem and that is that it there is no mainstream support; companies will not support it. New companies and middle-sized companies are willing to endorse new technologies that look good on the company and they are often the great innovators that decide the next decade.

Old and bigger companies want to downplay the technology as long as they can so that they can either keep their profits, switch over to the new technology while they can, or squash out the entire industry itself if they can't make that switch.

It's a repetitive pattern as all new companies either have to comply with what's currently being done or they have to fight the bigger companies, and this requires public approval. If the public does not support them, they will not win the fight and that means if blockchain is not supported by the mainstream public then blockchain will not be implemented in all the areas we know it can be because companies with mainstream public approval will be able to effectively fight it back.

Government Acceptance

Governments Are Slow

When it comes to practically anything, the government is slower than the slowest snail on the planet. Getting things done through the government takes years and sometimes even decades depending on the situation. Very rarely does a government conduct something that takes no less than six months because governments tend to spend their time arguing over the final details and ensuring that there isn't a vagueness to the law. This is highly beneficial because that means that laws can be challenged on specifics and then people from the public can challenge those laws if the specifics do not meet that government's own rules; a contradictory of terms.

The problem with which I'm pointing out here is the number of laws concerning blockchain because a great example is vaping from 2009. The American FCC and other regulatory bodies regrettably classified vaping or, rather, the products associated with vaping as tobacco related products. In a tobacco product such as a cigarette, one needs a way of lighting a fire and then the consumer will breathe in the smoke of the substance. In vaping, the coil that's wrapped around the cotton boils the electronic fluid otherwise known as e-liquid so that a vapor is produced that contains those chemicals. In the definition of science, the only two things that these two different categories of products have is that they are delivery systems and they deliver nicotine. In health, practice, and nearly everything else the two are completely different. Yet it took the government nearly a decade after the product hit the market to classify this substance and its delivery systems as tobacco-related, which many think is an incorrect classification. It took a decade to apply existing laws to a product. Imagine how long it would take for those various trust laws that are applicable to money and technology to be applied to cryptocurrency and the blockchain.

The Bane of Cryptocurrency on Blockchain

Those in the government are not exactly the brightest bulbs when it comes to technology because it's not their job. Weather hasn't been their job until the past two decades when it became extremely relevant. In other words, most governmental leaders don't know about technology because they really didn't grow up with it and it became a tool for their professional life that didn't require them to know the specifics. Professionals in the technology field handle those specifics for them and this is the problem that we face with blockchain in the governmental body.

Blockchain is confusing to entry-level programmers and yet programmers would have been able to ask better questions than Congress did of Mark Zuckerberg. So, to expect governmental bodies do understand the concept is pushing it. Time after time after time, it's been displayed that governmental entities that receive the limelight often do not have the intellectual capacity for medium to advanced technology. Hillary Clinton with her email server, the creation of a cyber division department within the government after it was a threat for three decades, and the multitude of other cyber security issues that have blasted news headings since the 1990s. People who have a political career, on the average, did not grow up with technology and tend to have very little experience with technology.

Difficult to Regulate

Not only that is but blockchain technology morphs every year. Bitcoin blockchain is different than Etherium blockchain and so on and so forth. How can you make a consistent set of federal laws that the blockchain and users of the blockchain can commit to if it changes every single time you have enough time to make a law about it? It is incredibly difficult to regulate the use of a blockchain and because blockchain updates as fast as a neurotic programmer can code a new application, it is nigh impossible to define the blockchain when the blockchain has variations.

The point of it being difficult to regulate is actually an internet-wide trend because many of the laws that are applicable to the internet in the country are not necessarily applicable to all countries. This is actually to the detriment of the user experience because what winds up happening is that companies, in an effort to easily conform to the law, use the most difficult to follow rules to set the guidelines on their platform. For instance, when the European Union came out with a new directive the giant websites of Google, Facebook, YouTube, Twitter and the like all updated their terms of service at the same to and it happened to be almost identical to that of the new directive. In other words, if you want to use YouTube you almost have to follow the exact same rules as those in Europe. International law has now superseded sovereign law via the Internet because it's easier for companies to conform to that.

Difficult to Monitor

The last part is the concept of being able to monitor the activity to ensure laws are being followed. Several times we have seen the government try to overstep its bounds and tell companies to lower their encryption standards because the government can't break it. What do you think is going to happen with blockchain technology?

The purpose of the government is to have control over its citizens for the most part. So when it comes to things like police officers and following the law, the government likes to claim that it needs an eye on everything to make sure that everything is being followed to the letter. Even in a country like America where it used to be the distance between homes that insured privacy, encryption ensures privacy online yet the government previously found it easier to have things like cities so that both houses and businesses could be located in the same spot just as it is easier to have encryption that isn't really difficult to break. There really is no excuse for the government asking for a way to decrypt something when that ability in it of itself defeats the purpose of encrypting. For example, if revolutionaries in the past that overthrew oppressive governments had those governments listen in on every conversation, those governments would not have been overthrown. In other words, encryption is another form of protecting the individual from the government and the government doesn't like that. So what's to say that the government, when it makes laws regulating the blockchain, won't try to build in a backdoor like they did with laws concerning technology products in workstations, they've been trying to do with smartphone companies, and consistently try with internet service providers?

Financial Institution Acceptance

Cryptocurrency Has Proven to Be Unstable

Once again, we have to tie this back to cryptocurrency because cryptocurrency and the blockchain have not been separated into their own entities in the eyes of the public. Just as I had mentioned before that negative perception of cryptocurrency bleeds into the perception of the blockchain, so too does the perception of cryptocurrency be unstable. Cryptocurrency is notorious for both jumping up in value and drastically decreasing in value for seemingly no reason at all. In most market trends, you can predict when a company might get a lot of money and might lose a lot of money, but the existence of the Dogecoin is proof enough that something that no one thought would have value has value in the cryptocurrency market.

Knowing about how unstable cryptocurrency has been, it is difficult to get someone to understand that while cryptocurrency uses the blockchain, cryptocurrency is not the blockchain. The technology industry has a huge problem with technophobia by people in power. People who have been in power for more than 50 years, whether it is commercial or governmental, very rarely understand how technology works and usually mistrust technology. It is so common in society that it has actually become a stereotype of anyone over a certain age. Yet, those people tend to be the true leaders of very important industries.

It Takes IT to Explain it

To make it even worse is that an IT professional usually has to explain what the differences are. This may not seem like a problem to IT professionals until you realize that the vast majority of businesses underestimate the importance of the IT position. There are countless examples of businesses hiring a freelance developer to build their website so that it looks and functions the way that they want it to. They then do not keep that freelancer on for maintaining and updating that software. This problem is getting better, but it is still bad. Most companies do not have a maintenance team, most of the companies that do have a maintenance team consider their maintenance team to be the IT support that a company like Squarespace or HostGator provides. The problem with a service like Squarespace is that even though they can provide security, they can also, potentially, make you more vulnerable because if they get hacked you get hacked. The only difference in that situation is that if they get hacked, every company associated with them also gets hacked, leading to tons of websites getting hacked all at once and the incentive to do so is much higher. It is a huge security flaw that most businesses that are hiring in the commercial food chain realize and is the reason why they tend to not use them. There are other reasons, but that is a linchpin reason.

Companies are hiring freelancers to create their own form of cryptocurrency because they know how much money can be made from it. When it takes an IT professional to explain it you have to consider the ignorance of the average consumer that's going to use it. Not everyone uses SolidWorks or AutoCAD yet the two programs are quintessential in our functioning society. Yet many people have only ever heard of one of them and have only ever heard it in passing. These are software that build the buildings they live in and build the parts that are in practically all of their products, yet they are names that would be lost on the vast majority of them. The blockchain is a very similar type of technology with the ability to replace most of how everything functions in society yet if you ever bring up the blockchain to anyone that was an average consumer and not in financial, marketing, business management, or a programmer they would understand you better if you talked about Bitcoin.

It Takes Cooperation

Perhaps the worst part about this is that you have to cooperate on multiple fronts for this technology to even have a purpose. For instance, in the car dealership example you have to have the car manufacturer to agree to use your system, you have to have the local governments agree to your system, you have to have the credit union agree to your system, you have to have your system inspected to ensure compliance with the existing laws, and you have to have a buyer willing to purchase from that ecosystem at the bare minimum. There's so much that you have to have cooperating with you that if any one of those fail it is almost unviable to actually implement a business based off of it.

Not only this but you also have to have a system that people can understand because if you leave it to a programmer to name things then only the programmer and the people who talk to the programmer would understand the purpose of those names. Therefore, you have to have a content designer and sales marketer that are also able to understand how the blockchain works in your ecosystem just to make the website understandable and functional. It is an incredibly enormous task to take on but that doesn't mean it's not worth doing it because the first person that does take on this issue and succeeds will find themselves as one of the top in their industry.

It Takes Mainstream

Last but not least is the news cycle. The news cycle is a bought Enterprise and I mean to say that those who own these news companies are often people in positions who own your competitors. What you will find is that if you try to go up against them, they will use everything in their power to give you as much bad PR as possible and trying to block every financial path that you have.

A good example of this is a company that was a startup known as Subscribe Star. Subscribe Star is a competitor of Patreon.

Patreon made an ideologically political move in many people's opinions against the Republican side in America and so some quickly looked to Subscribe Star as an alternative. Also, as soon as Subscribe Star got going, PayPal decided to quit its involvement with the company; almost as if there was some sort of communication between PayPal and Patreon. Therefore, the company had customers but have no way to receive the money essentially killing them as a company. As of writing this book the situation is still being handled and we will see what will come of this, but it serves as an example of what happens when these competitors try to target you.

Breaking the Mechanism

Users Are Still Lazy

The primary problem with securing information online is that users are incredibly lazy to a degree that would amaze most people. You see, user input, as much as it's needed in order to actually be the product, creates an encryption problem magnitude higher than most people understand. Most people understand that when you use a password several times, it's very easy for a person to figure out what your password likely is. It's the reason why there's an online list of the most commonly used passwords because, again, people are incredibly lazy. What happens is that if you know that a user goes to this website and then goes to another website but uses the same password or has a high likelihood of using the same password for both of them, you can actually figure out what the password is based off of which letters are different from each other. No matter how you encrypt something, if the password is the same you can create a mathematical differentiation between the two and effectively break both encryptions by having the same password in different encryption methods.

A huge problem with the entire blockchain being encrypted is that someone else can figure out how it's encrypted if you don't handle it correctly. You can figure out how it's encrypted in very much the same way that I just described to you because if a user's name is the same as the previous users name and the users name before that, what happens is that a whole bunch of users names will pile up that are the same and if a person has access to that they can then run it against names with the same amount of letters. The name was a central source of why the encryption was broken. Using the same encryption pattern will create an encryption that is identical in every single situation and then the person can reverse-engineer how you got to that point of encryption. It takes a significant amount of time to do this, but it's not anywhere near the level of Brute Force breaking an encryption should take.

Using the blockchain because it's an encryption method is an experimental technology because unless you are able to cordon off blockchains then the entire blockchain is open. When the entire blockchain is open you then risk the problem of someone getting access and then being able to figure it out based off of these factors. Essentially getting access to the usernames and passwords or perhaps the real identities of these people very quickly. Therefore, sensitive information can be leaked online almost as easily as someone making a backdoor into a database. The primary difference here is that unlike making a back door into a database, blockchain allows a user to create modular parts that can use different encryption patterns based on the module that it's in. What this does is it blurs the encryption method and makes it so that it is significantly more difficult to break the encryption. So, to give you an example, let's take those two names again and say that they are the same name but with different encryption methods. Unless you know specifically that those are two of the same names, the only way that you can figure out that they are the same names is the number of unchanging letters and the commonality of letters. The commonality of letters refers to the frequency in which we use the letters in our alphabet. I think the most commonly used letter is an e and that is because it is very common in everyday words but vowels, in general, are the most commonly used letters in the English alphabet. Knowing this, you can figure out which ones are vowels if there are enough letters in a name. There are ways around it but by using two different encryption methods, a direct comparison cannot be made and so the reverse engineering process would require brute-forcing both encryption methods rather than finding a mathematical differentiation between the two.

Encryption is a Delay Tactic

Having said that, a lot of people have this misconception that encryption is a permanent guard against people who want to get at sensitive information. These usually involve people who are outside of the tech industry and don't really think about how information is secured on a daily basis. In other words, the vast majority of society doesn't actually understand how information is regularly secured on different platforms.

Those that are in the tech industry understand that encryption is really a delay tactic and we are getting better at increasing the amount of delay that there is. For instance, the first most common form of encryption that we know of is actually known as rot13 and this is because Julius Caesar used it whenever he would send out messages that needed to be encrypted. Once you understood how this encryption method worked it would be really quick to figure out what the contents of a message was but until then no one had used encryption and so no one could figure out what messages meant.

Nowadays, the same thing can be done with current encryption methods and the only difference is in the amount of time it takes to do just that. Encryption is a method of hiding information by combining different elements together. Once you know what those elements are, it's a matter of running through the different combinations to figure out the exact combination that you used to encrypt all your stuff. Therefore, when it comes to encryption it's really just extending the amount of time that it takes a CPU to run through the combinations required to find the one combination that will be used to break your encryption pattern.

How to Overcome these

Educating the Public

Don't Primarily Focus on Adults

When it comes to educating the public a lot of people like to say that we need more young adults to learn this material so that it's quickly applicable. The problem is that a little bit too much is focused on teaching adults because adults are not generally going to experience the blockchain boom.

To understand this, you have to look at the smartphone because smartphones are really a relatively new concept. Starting in the year 2001 with Japan being the front-runner for them and we began to see them at around 2006 in the U.S. It took nearly a decade for everyone to really have a smartphone because what happens is that expensive technology is adopted by those who can afford it and then the popularity causes companies to create cheaper forms of it. This means that the blockchain, which is still a rather obscure technology, is still in the stages of the cell phone era.

Cell phones have been around for far longer than smartphones and it was because it didn't require much for cell phones to be created. It doesn't require much for a cryptocurrency and blockchain to be created. However, it did require a lot of expensive upfront cost for you to justify the need for a cell phone and as we progressed to the smartphone, we began to see cell phones at reasonable prices for the first few years before smartphones really hit the market. Once smartphones hit the market, cell phones were almost immediately made obsolete as a result of this.

Cryptocurrency started in the early 2010s with its invention going back much further, which represents a similar trend line in terms of creation. The concept and the first creation of cryptocurrencies was about two decades before we really saw something as big as Bitcoin. It was an academic idea for the most part, but once Bitcoin hit the market it began to accrue a lot of popularity once it had value and people knew it was easy to obtain. Now that Bitcoin has had a time to have its boom, what we are seeing now is that Bitcoin is slowly coming back down to its normal stabilized level. This is known as a crash for people because they didn't see it coming nor were they able to prepare for it. Blockchain technology is a specific component of cryptocurrency just as cell phones are a specific component of smartphone technology. It has taken a number of years to go from the blockchain as we knew it with Bitcoin to the Smart Contracts that we know with Etherium. Within about a decade we will see blockchain technology being used within companies and being almost as pervasive as smartphones have become now, but professional adults that could use this right now won't feel the effect until about a decade from now. This means that teaching young adults about this really teaches the 30-40-year olds how to get along with blockchain world.

Put It In Teen Products

To really have a decent effect, you need to affect the teen market. Cell phones did not expand as much as they did because of adults because adults tended to use cell phones as a utility and adults still tend to use smartphones as utilities. The ones that made it fashionable to have a new smartphone every year were the teens that grew up getting smartphones handed down to them from adults. After all, you don't exactly just want to throw away a $300 piece of hardware. Your teenager might get some use out of it so why not hand it down? Essentially, smartphones began to explode very quickly because the teenage population got to have their hands on them at an early age and then began to expect them as they became teenagers.

To have people understand what blockchain can do for you is a requirement of making the blockchain something that teens can use. Now, the average person will not understand what that means but considering the vast majority of encryption did messages are utilized by the teenage age group and young adults, you can begin to understand how to implement it with teenagers.

The reason why teenagers are so important is because teenagers are incredibly social. At that point they have a mixture of trying to get with their friends but also wanting to be able to afford their own items. Essentially, they are their own advertisement at that point in time and they are able to maintain a level of income that can usually be spent on expensive products. Cryptocurrency was a very popular tagline in YouTube videos a few short years ago because there was a lot of interest in generating money with cryptocurrency, not with adults but with teenagers and young adults. This is because teenagers and young adults had money to spend, to waste, which meant that a lot of investment went in to cryptocurrency with these teenagers and young adults. This was a form of making teenagers open to the idea of cryptocurrency and now we have an entire generation that has seen the benefit of cryptocurrency and the potential problems with cryptocurrency.

308

The reason why this is important is because teenagers will be the ones that fix the problems of the past so any problems that teenagers experience with cryptocurrency now, teenagers will be able to fix when they enter the job market. Introducing the blockchain to teenagers introduces them to the benefits and the problems of blockchain technology, which represents an opportunity to implement blockchain into an expected form of life. By doing this, blockchain can very quickly obtain popularity and products that utilize blockchain will be popular amongst teenagers. By advertising and educating teenagers, companies will be forced into implementing blockchain technology as a way to compete with other companies that are utilizing the teenage fascination with blockchain technology. This has actually been seen in marketing advertisements that use the word gamer.

They use the word gamer because for some reason the marketers see a 20% increase when the word is used, regardless of whether that keyboard is the same as a regular keyboard. You can have a keyboard that is generic and then you could have a keyboard that's labeled gamer and the one that has the gamer label will do 20% better. This is because teenagers don't really think too hard about the specifics of a product, they think a lot about The Branding and what it means to have that product. Therefore, while someone might think that Logitech is the best keyboard on the planet because it's got a smart screen built into it, a Chinese manufacturer who comes out with an RGB keyboard that took maybe $10 to make can sell more because it's associated with the brand of gamers and being a gamer is something Logitech doesn't advertise. This makes it relatable to teenagers and with teenagers, the most crucial selling aspect is how it relates to the teenager's life. By teaching teenagers what blockchain technology is and how it can be used in their life, you effectively motivate the market to develop better more complicated forms of blockchain to increase the benefits that blockchain can provide.

Teach the Kids

To go even further than that you need to teach the kids because the kids will be the ones who become teenagers and see the problems that the teenagers cause. The best way to understand this is the gaming industry itself. If we were to rewind by about 21 years, we will be introduced to what most considered to be the best beginning to the gaming area that there was and that was the late 1990s. During the late 1990s, you had consoles that were coming out like the PS2 and the Xbox and games that had stayed with kids during that time until now. As time has progressed, those games have gotten more graphically and programmatically more complex and better in some cases. Storytelling has pretty much stayed the same and it's really just the graphical details that have changed along with how much you can interact with. Those systems inspired the kids who are making games now, games like Red Dead Redemption 2 and Spider-Man and those are some of the highest selling games of the year this book is written in. The problem is that the AAA industry has a really bad reputation at the moment because somewhere along the way, smartphones were invented and free games had microtransactions in them to make profit off of free games. If you think about how microtransactions are the bane of current video game technology in terms of making money, you can actually think microtransactions are really good because they tend to make a lot of money.

The problem with that way of thinking is that you sacrifice the enjoyment of video games with every microtransaction you shove into your game. These kids who are playing games as adults experienced a time where there were no microtransactions. They make games for the story and the enjoyment of video games, which means the Indie Market that was small many years ago has now exploded with video game makers who grew up during that time.

Along the way, video games became a form of making money rather than an art form. Video game companies that were making the biggest games playable, the ones developed on the PS2, stayed around and began incorporating people who grew up with microtransactions inside of free games. These same people have company leaders who try to exploit every single way of making money possible yet now we are seeing trends in video games in the AAA gaming sometimes doing worse than Indie Games. A great example of this is Five Nights at Freddy's, which made some of the highest game sales of all time with Indie Games and competed with a lot of AAA industry games. So you have a cross between those that grew up poor wanting to make the most money and those that grew up with the vision of beautiful and impactful video games crossing paths within the same industry.

Teenagers will use blockchain technology in order to do certain things that they're probably not supposed to do and make transactions with people in other countries as we come online more often. Kids will look at how teenagers are making use of the blockchain technology and they will find something wrong with that use, which further opens up the possibilities of what blockchain can do. Once kids and teenagers understand what blockchain technology is and purchase more products with it, you will see companies quickly conform to using blockchain in every aspect of their company that they can benefit for themselves but they will also realize the marketing benefits that come with saying that their technology is what the teenagers and kids want.

Working within The Loopholes

Open Companies in Digitally Unregulated Areas

Facebook and YouTube are fantastic examples of companies opening businesses in areas that are not normally heavily regulated. Facebook was a social interaction platform that allowed people to connect to each other long before it became the tangled web that it is now. YouTube used to be a very small video website that had incredibly small videos on it with a small percentage of society even knowing it existed. That was almost a decade ago and now they are among the biggest companies in the world while just now facing the curtails of regulation in the works.

You see, most communication when it came to forums and instant messengers went unregulated because it was actually pretty small and fell under the guise of free speech. For instance, let's say that you wanted to start a forum for everyone who really liked to build printed circuit boards. The only way that someone could be allowed into the group is if you invited them into the group and then, as the Forum owner, you could kick people out you didn't want in there and they couldn't do anything against you. Now, a lot of people don't realize that this is essentially the concept of Facebook and Twitter. Facebook and Twitter are two online forums that are extremely convenient and so a lot of people use them. Twitter is an open forum for the most part and recently started adding the ability to privatize your messages so that your personalized forum could only be seen by people who you approved to follow you. Facebook has been a closed forum for a very long time with public posting, which meant you could have private messages while also being able to post to the public.

Apply Blockchain in Less Known Industries

The Raspberry Pi is an excellent invention of ingenuity and planning as it serves as a device for various purposes. It was the original single board computer that provided a lot of uses and gained a lot of popularity. The mainstream public didn't know about the Raspberry Pi for a very long time and the first way that it was actually handed out was when you bought a magazine. Over time, it evolved into a usable single-board computer for everyday tasks except for gaming and hardware intensive applications like Adobe Premiere Pro. You could use it as a document editor, programming machine, or anything that just required low-level computing power.

The reason why I bring up the Raspberry Pi is because it's a lot like the blockchain. The blockchain started out as a very simple idea and has had its evolutions over the past few years. The Raspberry Pi is practically known to most parents that are not minimum wage because it's a fantastic gift and project depending on how you want to handle it and it doesn't require a lot of experience. The blockchain, on the other hand, is still rather unknown to people and is not exactly user-friendly when it comes to explaining it. The Raspberry Pi represents a very lesser-known industry because the Raspberry Pi is not the only single board computer. Sure, it might have been the first, but it definitely isn't the only one that there is. In fact, there are a number of single board computers that are as powerful if not more powerful than the current release of Raspberry Pi. The only difference is that the Raspberry Pi had released in a lesser-known industry at the time and paved the way for the rest of these models. The blockchain can do the very same thing if it's given the opportunity.

For instance, why not develop a library system that bases itself off of the blockchain? Essentially, you just put an RFID chip inside of each book and it checks books in and out based on what's on the RFID chip and in the blockchain. Everyone knows what a library is but not many people actually know how the systems in the library function which means that they could be a useful way of expanding the control for blockchain. It would also give them a much better PR personality versus the criminal and unstable vision that cryptocurrency gives it.

Blockchain Implementation Via Creeping

That's the problem with today's generation, they expect change to happen overnight because they've seen it before. They've seen Lindsey Stirling up on the stage singing her heart out only to fail within a few months on the show she was singing for, yet she becomes an international star during the same time span. Justin Bieber was a similar story and this trend happens more often now than it ever did because once something is on the internet, it can spread like wildfire.

The problem is that this is not how new technology should be introduced. Most of everyone, as of writing this book, thinks that the world of VR is currently dead but it's not and it's actually expanding, it's just expanding at a very slow rate. Blockchain should be a technology that expands slowly because the slower something moves; the less people notice it moving at all. This means that eyes will not be placed on the blockchain until the blockchain has a secure foothold in society.

Believe it or not, this is how coding libraries work. A few years ago, no one knew what bootstrap was, a few years ago no one knew what react.js was, but they slowly made their way into the eyes of programmers and it became an adopted norm. Frameworks and Technology that's not seen as a good change by those competing with it should move at a very slow pace because if it moves fast then it won't be able to handle the challenges.

This means that the best way to really spread blockchain is to find some mundane use that it can have inside of your business. Perhaps you might want to create an instant messaging app that everyone at the business can use to communicate with each other and you just so happened to build it on the blockchain. Perhaps you want to clock the break times that people have and you build that on blockchain. Building small things out of blockchains will cause a very slow overtake of the current system that's in place. As it is said, a crab that is slowly brought to boil will not know that it is dying.

Setting Up the Finance Details and Security

A lot of the software that the finance industry uses is pretty much capitalized on because finance doesn't really change that much. I mean, a mortgage is still a mortgage, a loan is still a loan, buy and sell options are still buy and sell options, and the problem is that it's so slow in updating as an industry as a whole.

That isn't to say that Finance isn't a workable field and that isn't to say that there aren't updates made to the field, but the software around what a finance person does is pretty standard and doesn't really expand a whole lot because while there's an almost limitless product line money is still money.

Having said that, Finance deals in a lot of information that's sensitive. A company checking account number as well as the password to get in said checking account is not really something a company wants to share with other people. When someone goes about making changes to company accounts, this is yet another form of finance. A huge problem in company finance is determining how much money has been spent and the blockchain technology can actually help mitigate situations in which employees spend money on the company's behalf without the company's permission. You see, the way the situation normally goes is that an employee gets a card to spend it on certain items like gas but then they will accompany those items with food or drink at gas stations or even have a payout at the gas station. The way they usually get caught is at the end of the year when they're required to put in their receipt or at the end of the week, it's kind of obvious that one of the receipts is not going to be like the others.

The problem is that this could actually be stopped before it happens with an authorization card. An authorization card is a middleman of sorts but digitally and this refers to a card that you swipe at a credit card reader, but before the transaction goes through it is actually sent to a separate station for approval before the transaction is allowed. What occurs is that the authorization card requests the receipt to be transmitted digitally and then the person on the other end can approve or deny a transaction based upon what they read in the receipt. In other words, it would stop employees from stealing from the company on the spot. The employee would then be responsible for paying for the items that they tried to use the company money on.

How does this involve the blockchain though? Well, the same authorization card can be used to automatically collect the receipts of employees that you can trust (or your own transactions). The time, the place, and similar things can be stored in an authorization request and you would want to keep this information and easily something that you want to have an easy look up for. It is very difficult to navigate a database if there's no way to make sure that it's segregated so you could separate each transaction by company, which would effectively ensure that known restaurants or manufacturers could proceed with the transaction without having to wait for an approval, while gas cards could be cordoned off into their own section. This would keep a better record of all transactions instead of receipts, prevent company theft/embezzlement, and help to ensure the accuracy and speed of tax specialists.

Conclusion

Blockchain Is A Powerful Technology

Blockchain is a really powerful technology and can be used for quite a lot, but the problem isn't that it's a powerful technology but a difficult to understand technology. Oftentimes, the blockchain itself needs a very lengthy article or book in order to describe its power and its uses. However, powerful technology is still powerful technology, by explaining it we only increase the amount of access people have to the power of this technology. It is very important to understand that with powerful technology does come a greater amount of responsibility for ensuring that everyone knows the truth about it because what happens is that people fearmonger very easily.

Fear-mongering has been a part of human Society for as long as recorded history, from the Kraken to the Devilry in Books. This has never ended, and it happens with everything that's new. Everything that's new is called a detriment to society until everyone has it and then everyone thinks it's weird that it was called a detriment to society. Books were exclaimed to be the worst thing one could do to oneself and now the biggest Business Leaders in the world are exclaiming that they read at least one book a day. Right now, the digital age is upon us and the time in which we figure out what we're going to do with the internet now that it is a global society is considered to be the current detriment to society. People will get along more if they would just get off the internet except that the internet is no different than talking to a random stranger. People will get along more if video games stop being so violent; except that terrorism has always existed and that it hasn't really increased or decreased in the wake of video games. There's always science and logic to beat these down but fear-mongering always exists and so it is the responsibility of those that love this technology, this new invention to spread that explanation across the world so that we as a society can improve.

A World Devoid of Middlemen

This isn't about the blockchain, I mean it really is about the blockchain, but it isn't that the blockchain will solve all the problems it will just solve the middleman problem. With a blockchain you no longer need a guy at a dealership to sell you a car, you no longer need a cashier at the register to check to see if you purchased that item in your hands, you no longer need someone to check identification at the bar, and you no longer need to worry about whether your shipment arrived or not. The middleman is a specialized industry that simply exists as a means of making sure that we trust one another and considering that the internet is a global marketplace, trust should be as digital as the internet itself. We should not have to rely on another person to decide we trust information but rather the mathematical proofs that have always served science well.